Springer Theses

Recognizing Outstanding Ph.D. Research

Aims and Scope

The series "Springer Theses" brings together a selection of the very best Ph.D. theses from around the world and across the physical sciences. Nominated and endorsed by two recognized specialists, each published volume has been selected for its scientific excellence and the high impact of its contents for the pertinent field of research. For greater accessibility to non-specialists, the published versions include an extended introduction, as well as a foreword by the student's supervisor explaining the special relevance of the work for the field. As a whole, the series will provide a valuable resource both for newcomers to the research fields described, and for other scientists seeking detailed background information on special questions. Finally, it provides an accredited documentation of the valuable contributions made by today's younger generation of scientists.

Theses are accepted into the series by invited nomination only and must fulfill all of the following criteria

- They must be written in good English.
- The topic should fall within the confines of Chemistry, Physics, Earth Sciences and related interdisciplinary fields such as Materials, Nanoscience, Chemical Engineering, Complex Systems and Biophysics.
- The work reported in the thesis must represent a significant scientific advance.
- If the thesis includes previously published material, permission to reproduce this must be gained from the respective copyright holder.
- They must have been examined and passed during the 12 months prior to nomination.
- Each thesis should include a foreword by the supervisor outlining the significance of its content.
- The theses should have a clearly defined structure including an introduction accessible to scientists not expert in that particular field.

Huibin Wei

Studying Cell Metabolism and Cell Interactions Using Microfluidic Devices Coupled with Mass Spectrometry

Doctoral Thesis accepted by
Tsinghua University, China

 Springer

Author
Dr. Huibin Wei
Beijing Research Institute of Chemical
   Industry, SINOPEC
Beijing
China

Supervisor
Prof. Jin-Ming Lin
Department of Chemistry
Tsinghua University
Beijing
China

ISSN 2190-5053   ISSN 2190-5061 (electronic)
ISBN 978-3-642-43086-2   ISBN 978-3-642-32359-1 (eBook)
DOI 10.1007/978-3-642-32359-1
Springer Heidelberg New York Dordrecht London

Printed on acid-free paper

Springer is part of Springer Science+Business Media (www.springer.com)

Parts of this thesis have been published in the following journal articles:

Huibin Wei, Haifang Li, Sifeng Mao, and Jin-Ming Lin. **Cell signaling analysis by mass spectrometry under co-culture conditions on an integrated microfluidic device**. *Anal. Chem.* (2011), 83 (24), 9306–9313.

Huibin Wei, Bor-han Chueh, Huiling Wu, Eric Hall, Cheuk-wing Li, Jin-Ming Lin and Richard N. Zare. **Particle sorting using a porous membrane in a microfluidic device**. *Lab on Chip* (2011), 11, 238–245.

Huibin Wei, Haifang Li, Dan Gao, Jin-Ming Lin. **Multi-channel microfluidic devices combined with electrospray ionization quadrupole time-of-flight mass spectrometry applied to the monitoring of glutamate release from neuronal cells**. *Analyst* (2010), 135, 2043–2050.

Huibin Wei, Haifang Li, Jin-Ming Lin. **Analysis of herbicides on a single C30 bead via a microfluidic device combined with electrospray ionization quadrupole time-of-flight mass spectrometer**. *J. Chromatogr. A.* (2009), 1216, 9134–9142.

Sifeng Mao, Dan Gao, Wu Liu, Huibin Wei, and Jin-Ming Lin, **Imitation of drug metabolism in human liver and cytotoxicity assay using a microfluidic device coupled to mass spectrometric detection**, *Lab Chip* (2012), 12(1), 219–226.

Bor-han Chueh, Cheuk-Wing Li, Huiling Wu, Michelle Davison, Huibin Wei, Devaki Bhaya, Richard N. Zare. **Whole gene amplification and protein separation from a few cells**. *Anal. Biochem.* (2011), 411, 64–70.

Ling Lin, Hui Chen, Huibin Wei, Feng Wang, Jin-Ming Lin, **On-chip sample pretreatment using porous polymer monolithic column as solid-phase microextraction and chemiluminescence determination of catechins in green tea**, *Analyst* (2011), 136(20), 4260–4267.

Wu Liu, Huibin Wei, Zhen Lin, Sifeng Mao, Jin-Ming Lin, **Rare cell chemiluminescence detection based on aptamer-specific capture in microfluidic channels**, *Biosensors and Bioelectronics* (2011), 28, 438–442.

Ling Lin, Zhaoxin Gao, Huibin Wei, Haifang Li, Feng Wang, Jin-Ming Lin, **Fabrication of a gel particle array in a microfluidic device for bioassays of protein and glucose in human urine samples**, *Biomicrofluidics* (2011), 5, 034112, 10 pages.

Dan Gao, Huibin Wei, Guang-Sheng Guo, Jin-Ming Lin. **Microfluidic cell culture and metabolism detection with electrospray ionization quadrupole time-of-flight mass spectrometer**. *Anal. Chem.* (2010), 82, 5679–5685.

Dan Gao, Jiangjiang Liu, Huibin Wei, Hai-Fang Li, Guang-Sheng Guo, Jin-Ming Lin. **A microfluidic approach for anticancer drug analysis based on hydrogel encapsulated tumor cells**. *Analytica Chimica Acta* (2010), 665, 7–14.

Supervisor's Foreword

Nowadays life sciences are a very important part of natural science. The cell is the basic unit of an organism, from which we endeavor to crack the code of the living entity. Microfluidic technology, which currently is being developed rapidly, is widely used in the cell research field because of its microscale channels and flexible design. In recent years, many researchers have focused on studying cell secretions, intercellular structure, and signal transmission between cells based on microfluidic devices. Several detection approaches are applied to observing and determining cell activities. However, most of the approaches are restricted to imaging and tracing the target substance only, lacking a technique that could identify and analyze the molecules that act as essential factors in cell activities.

The main aim of this Ph.D. thesis is to develop an innovative cell analysis platform that can not only monitor the molecular target but also identify and analyze the potential factors. Cells are manipulated and captured on the microfluidic devices, and mass spectrometry (MS) is employed for qualitative and quantitative detection. Our key issue is connecting the microchips with MS, realizing cell culture and drug stimulation in the microchannels, and purifying cell secretions before their detection by MS.

The research was carried out not only at the Chemistry Department of Tsinghua University, but also with the cooperation of the Chemistry Department of Stanford University, to solve the cell sorting problems that we met in handling real biological samples. This thesis illustrates how to establish a platform coupling a microfluidic device with mass spectrometry and apply it to cell analysis. It also explores the possibility of using this analysis platform for drug screening and early diagnosis of certain diseases.

Beijing, January 2012   Jin-Ming Lin

Acknowledgments

I first express my gratitude to my supervisor, Prof. Jin-Ming Lin. His conscientious and diligent academic attitude will be my example throughout my lifetime.

I would like to thank Dr. Haifang Li for her valuable advice and generous help on my projects, as well as contributions to the published papers. I am grateful to the students and colleagues with whom I have worked in the laboratory: Jiangjiang Liu, Danning Wu, Zhaoxin Gao, Yuanfeng Pang, Chuansen Liu, Ting Fei, Dan Gao, and the others. I have wonderful memories of this hardworking and warm group.

I also would like to thank Prof. Richard N. Zare at the Chemistry Department of Stanford University. During the 9-month cooperation in Zarelab, he and the other lab members gave me enthusiastic support and help. I also thank the China Scholarship Council for the funding from my scholarship.

I also extend my thanks to my parents for their continuous love and encouragement. I thank my husband Dr. Xaver Karsunke, who has supported me with patience throughout the hard times, holding my hand while I was going forward.

Finally I thank the National Nature Science Foundation of China for supporting the projects.

Contents

5 Cell Co-culture and Signaling Analysis Based on Microfluidic Devices Coupling with ESI-Q-TOF MS

Abbreviations

μTAS	Micro-total analysis systems
PDMS	Polydimethylsiloxane
ESI	Electron spray ionization
MALDI	Matrix assisted laser desorption ionization
Q	Quadrupole
TOF	Time of flight
ESI-Q-TOF	Electron spray ionization-quadrupole-time of flight
MS	Mass spectrometry
TIC	Total ion chromatogram
EIC	Extracted ion chromatogram
SPE	Solid-phase extraction
PEG	Poly(ethylene glycol)
PEG-DA	Poly(ethylene glycol) diacrylate
SU-8	A series of negative photoresist
SPR	A series of positive photoresist
AZ	A series of positive photoresist
CCD	Charge coupled device
PBS	Phosphate buffered saline
BSA	Bovine serum albumin
PCR	Polymerase chain reaction
RSD	Relative standard deviation
LOD	Limit of detection
LOQ	Limit of quantification
PS	Polystyrene
DMEM	Dulbecco's modified eagle's medium
RPMI	Roswell park memorial institute
PC12	A cell line derived from a pheochromocytoma of the rat adrenal medulla
GH3	Rat pituitary tumor cells
HepG2	Human hepatocellular carcinoma
REH	Human leukemia

ECM	Extra cellular matrix
PLL	Poly-L-lysine
rpm	Revolutions per minute
dpi	Dots per inch

Chapter 1
Introduction

1.1 Background Information

Microfluidic technology, also called "lab on a chip" (LOC), miniaturized the basic units of biological, chemical and medical laboratories using a chip with a size of only several square centimeters. This technology is rapidly developing in recent years. The manipulations of sample preparation, reaction, separation, and detection were integrated into micro-scale channels, in order to achieve a portable, automatic, rapid, and accurate analysis system. The concept of micro-total analysis systems (μTAS) was first defined by Manz [1] in 1990. During the last 30 years, micro-fabricating techniques developed rapidly, as well as the separation and detection methods. Thus, the microfluidic devices fabrication was greatly improved. Micro-valves [2] and micro-reactors [3] were successfully integrated in the microfluidic devices, which provided the essential conditions for the integration and automation of microfluidic devices. As a fast developing analysis technique, μTAS was widely applied in various research fields, particularly in disease diagnosis, environment monitoring, immunoassays, and protein research [4].

Cells are the basic units of living species. All organ and tissue reactions are essentially the behavior of each single cell, which builds up the organ or tissue. Based on the research of single cells, the principles of cellular functions and their interaction could be better explained. In recent years, the attention paid on cell research covered the fields of respiration, photosynthesis, signal transmission, and cross membrane transport. The research objectives reach from the cell mass, single cells, subcellular structures, down to the molecular structure. Recently microfluidic devices are more and more widely applied in cell research. Not only because they have micro-scale channels with similar dimensions as cells, but also have advantages such as high separation efficiency, fast analysis process, highly flexible design, little sample pretreatment, and they can be highly automated. Due to the small dimension of the microchannels, the specific surface is quite large, and the reagent

H. Wei, *Studying Cell Metabolism and Cell Interactions Using Microfluidic Devices Coupled with Mass Spectrometry*, Springer Theses, DOI 10.1007/978-3-642-32359-1_1,
© Springer-Verlag Berlin Heidelberg 2013

consumption is very low. For these reasons, microfluidic devices are integrated into portable instruments for point-of-care diagnosis and medical care, which most of the traditional analytical techniques cannot achieve.

The cell research based on microfluidic devices is emerging in two directions: single cell analysis and signal transmission. In traditional analysis theories, it was desired as well as assumed that cell cultures were of a homogeneous nature, and that analyzing a group of cells would give an accurate assessment of the behavior of the cells in the respective culture or tissue. The average response of the cells was interpreted as the response of all cells in that sample. However, the average cell behavior normally covered and neglected certain exceptional cell behaviors due to the difficulties in monitoring such detailed phenomena. For example at the early stage of cancer, only very few cells carry the information of organ malfunction. The inspection through the organ with traditional medical tests cannot discover the pathological issue at the early stage. Furthermore the concentration of cancer marker collected from a blood sample is lower than the detection limit of traditional analysis methods. The ignorance of the organ malfunction at the early stage results in the detainment of early diagnosis [5]. The high-throughput selection and data collection for single cells in microfluidic devices are making efforts on the development of new technologies for the early diagnosis of diseases.

The dynamic study of signal transmission between living cells can increase the understanding of the interconnecting molecular events continually taking place in each cell. Furthermore the signal transmission between single cells can explain the connection and interaction behaviors between tissues and organs [6–8]. Based on the highly flexible designs of microfluidic devices, the cells can be co-cultured in a desired micro-environment. These are able to be controlled and changed easily, and the signal factors between different cells can be studied via proper detection approaches. Currently, there is an increasing awareness of these issues, especially in the fields of cell sorting, co-culture, manipulation, separation and detection methods.

1.2 Microfluidic Devices for Cell Analysis

Due to their unique characteristics, cell analysis techniques based on microfluidic devices have been greatly developed, including the microchip fabrication and functional unit integration. Cell culture [9–13], sorting [14–18], and lysis [19] were achieved on microfluidic devices. The unique advantages of microfluidic devices determine the wide application prospect on cell analysis:

1. Micro-scale channels and reactors are able to be integrated in the devices, which have the same order of magnitude of cells, and provide the possibility to manipulate single cells.
2. Multiple materials could be selected for microfluidic device fabrication. Especially it is possible to use biocompatible materials to build the micro-environment for cell cultures, which mimic the *in vivo* environment in living species.

3. Integrated micro-reactor greatly reduce the consumption of reagents, which ensures an energy-saving and environmental-friendly process.
4. Compared to traditional analysis methods, time consumption is also decreased, since multiple functions could be realized on a single chip.

1.2.1 Materials for Fabricating Microfluidic Devices

Three requirements have to be met for materials used to fabricate microfluidic devices for cell analysis: non-toxicity and good biocompatibility, transparent for observation and detection, and easy surface modification. Glass and polymers are the materials most commonly used for fabricating microfluidic chips. Glass chips are found of good optical properties, and with a surface which is quite easy to be modified by biocompatible chemicals. Disadvantages are that the fabrication process is complicated, and a stable bonding between two glass chips is hard to be achieved. For those reasons glass is normally combined with other materials for microchip fabrication. The advantage of polymer chips is the low cost, and that rapid casting fabrication is possible. Plus the chemical inertness and thermal stability, polymer chips are the first choice for cell analysis.

Due to its porous character which is important to provide an essential air environment for the cell culture, the most commonly used material is polydimethyl-siloxane (PDMS) [20, 21]. Furthermore PDMS is transparent, flexible, chemical inert, and biocompatible. Ingenious approaches of cell analysis were already explored on PDMS microfluidic devices. Moreover, there are various ways for bonding PDMS chips: Simple physical adhesion could be employed to realize the reversible linkage, and by chemical modification irreversible linkage could be achieved. A disadvantage of PDMS chips is that the hydrophobic and porous surface may cause the diffusion of hydrophobic compounds into the microchip structure, which can lead to concentration changes of hydrophobic compounds in the microreactor. Regional modification was adopted to prevent the concentration change in PDMS microreactors [22].

Polymethylmethacrylate (PMMA) [23], polycarbonate (PC), polystyrene (PS), and polyethylene terephthalate (PETG) [24] were used to fabricate microfluidic chips for cell research. Poly (dl-lactic-co-glycolic) (PLGA) [25], and poly (glycol succinate) (PGS) [26] are also employed as biodegradable support materials because of their perfect biocompatibility and plasticity, in order to mimic the *in vivo* three-dimensional environment in organisms.

Culturing cells in three-dimensional hydrogel was found to help cells retain their native behavior and differentiated state. In particular, due to their uniform property, good permeability for small molecules and similar physical features with the *in vivo* matrix, the potential significance of hydrogel cell cultures in microfluidic devices has gained increasing attention. Commonly used hydrogels are polyethylene glycol hydrogel (PEG), collagen, agarose, etc. Heo [27] and co-workers first introduced PEG in cell cultures. Embedded in hydrogel microstructures, *E. coli* could be

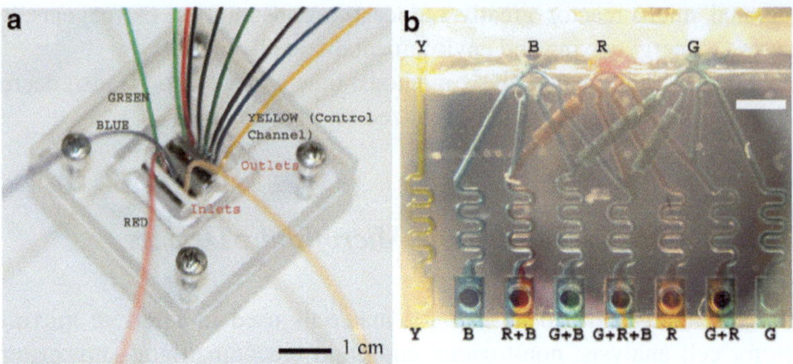

Fig. 1.1 Three dimensional cell culture microfluidic device fabricated by combined materials. (**a**) Picture of a three dimensional cell culture microfluidic device. (**b**) Mixing process of three different solutions (Reprinted from Ref. [29], with permission from Elsevier)

observed in a dynamic monitoring. Ling et al. [28] used a mixture of cells with agarose as a material for microchip fabrication. The stimulation chemicals were injected through the microchannels formed under the cells, and the cells dynamic response was observed.

More options could be selected by combing different materials. Multiple potential choices were developed, such as combing glass chip with PDMS, fabricating hydrogel microstructures in PDMS chips, and integrating different polymer materials. Figure 1.1 shows a three dimensional cell culture microfluidic device fabricated by Liu et al. [29], chosing poly(chloro-p-xylylene) combined with PDMS. In this way it was possible to evaluate the effect of three different drugs and their combinations on single cells in one experiment.

1.2.2 Flow Control in Microfluidic Devices

The key technology for controlling the process in each chamber is the flow control, which connects different microstructures. Highly integrated and automated microfluidic devices, containing hundreds of chambers, require micropumps and microvalves to accurately drive or stop the flow. Especially in single cell analysis, the processes such as the nutrition transmission to the cell culture chambers and sequential stimulation by drugs with gradient concentrations, require highly accurate manipulations. The proper connections between the chambers for cell capture, cell culture, cell stimulation, and secretion pretreatment, which also need to prevent interferences, are important for cell analysis in microfluidic devices.

Active driving forces were the most common choice in flow control. Peristaltic pumps, injection pumps, and vacuum pumps are normally adopted in laboratories to

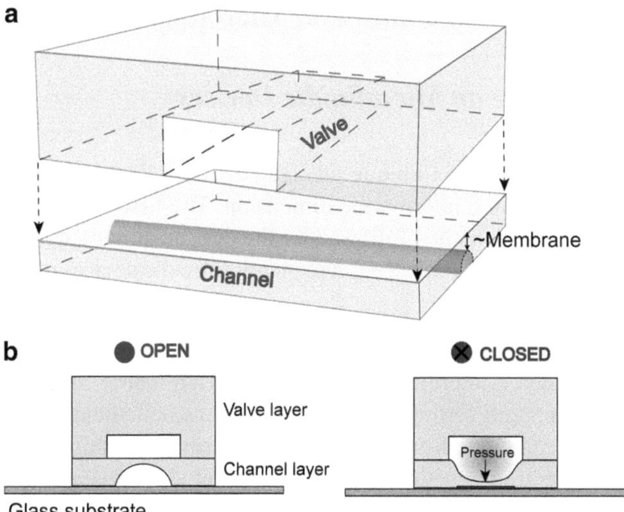

Fig. 1.2 Principle scheme of fabricating the microvalve by multiple layers of PDMS microfluidic devices. (**a**) 3D view of valve, channel, and membrane layers is shown. (**b**) The microvalve is actuated by increasing the pressure of the fluid inside the valve layer

drive the flow in microfluidic devices. Self-organized bacteria were introduced into microchannels with an active flow driving force by Kim et al. [30]. A continuous flow with a speed of 25 μm/s was created and maintained for several hours. The flow rate was able to be regulated by changing the geometry of microchannels, or replacing the reagents to stimulate the bacteria. The passive driving was also selected in some cases. Meyvantsson [31] and co-workers used the surface tension instead of the traditional manual injection to drive the flow for refreshing the cell culture medium. In this way two types of cells HMT-3522 S and Hs578 S1 were successfully cultured for more than 4 days.

Several types of microvalves were developed to control the switching status of the microchannels. The most important microvalve type was designed based on the flexibility of PDMS. Unger [32] and co-workers fabricated microchannels with a curved cross-section by reflowing positive photoresist. A very thin upper channel wall was constructed, and pressure was applied from the top to close the microchannel to cut off the flow (Fig. 1.2). Additionally, phase transition valves, pneumatic valves, electric valves, and solenoid valves were also widely adopted in microfluidic devices [33] to achieve the precise control of flows in microchannels. The highly integration of a large number of microvalves, which could be automatically controlled by a software on a computer, provides the essential conditions for the simultaneous stimulation and observation of large single cell arrays. King [34] and co-workers applied inflammatory cytokines with different concentrations on high throughput cell arrays by a computer controlled valve system, in order to study the cells gene expression.

1.3 Cell Capture and Culture on Microfluidic Devices

1.3.1 Cell Capture on Microfluidic Devices

Microfluidic devices revealed unique advantages for cells manipulation, because their channels are within a micro-scale dimension. The first step in cell analysis is to trap and fix the cell, which is followed by the desired stimulation and detection. Current developed methods for cell capture on microfluidic devices are as follows:

1. Magnetic force: The specific binding between cells and magnetic particles were employed to manipulate the cells in a magnetic field. Berger [35] and co-workers developed a microfluidic device with asymmetric obstacles inside the microchannels for cell sorting while applying a magnetic field, which was based on the principle of the Brownian motion. Microfluidic chips were combined with a micro-electromagnetic array by Lee [36] et al., to generate a controllable magnetic field to manipulate a single cell in the flow. Furdui [37] and co-workers trapped and fixed cells by a magnetic field, to capture and enrich rare target cells in blood. The method of cell capture by magnetic force is a clean and gentle approach, but new materials for producing magnetic particles and novel technology for coating the particles are required to meet more demands for capturing a larger variety of different kinds of cells.

2. Optical tweezers: This is a non-contact and non-contamination method for cell manipulation. Biological molecules could be fixed on insulating particles by optical tweezers, and then the particles were captured in an electric field gradient. Umehara [38] et al. trapped single bacteria in a microarray for culturing by using optical tweezers. Each individual bacterium was isolated and fixed in the microchambers, and different culture mediums were introduced to study the behaviors of homologous bacteria. Enger [39] and co-workers employed the optical tweezers to move *E. coli* to the target location within seconds, and the cells response to different fluorescent markers were studied. Although the single cell was accurately controlled by optical tweezers, this approach requires highly precise equipment for optical positioning. Its application is restricted by the complex optical equipment, professional manipulation process, and high equipment costs.

3. Mechanical force: Fluid dynamics, gravity, capillary force, wetting, and adhesion force were mainly employed to trap and manipulate cells. Most of the works focused on fabricating specific microstructured arrays, which just fit the size of a single cell, in order to capture the individual cells and obtain the single cell array. The following examples give an overview of the use of mechanical forces for the cell capture (Fig. 1.3): Microarrays were fabricated in the microchannels along the horizontal direction of the flow and the cells were accurately trapped in each microstructure [40–43]. The U-type or hat-shape dams [44] were fabricated to obtain a precise single cell array. Similar structures were constructed to achieve the accurate matching between two different types of single cells, which contributed to the study of cell fusion [45] and cell co-culture [46].

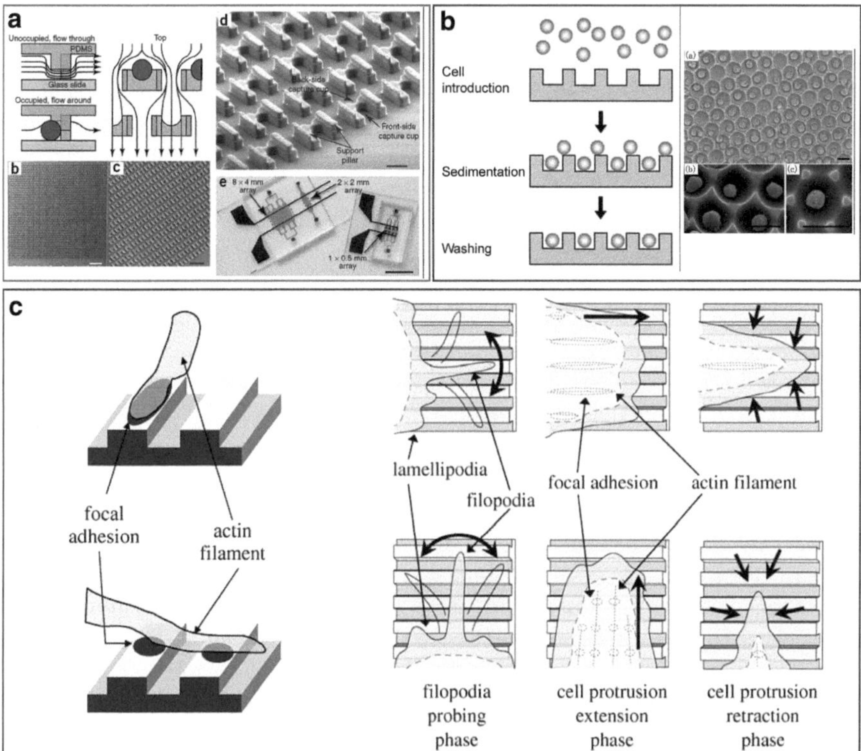

Fig. 1.3 Cells capture by mechanical force. (**a**) Single cells were trapped by the fluid dynamics (Reprinted by permission from Macmillan Publishers Ltd: Ref. [45]. Copyright 2009). (**b**) The gravity was employed to introduce the cells into the microwell array (Adapted with permission from Ref. [47]. Copyright 2010 American Chemical Society). (**c**) Cell alignment on the nanogrooved substrate (Reprinted from Ref. [53], with permission from Elsevier)

Microwells were fabricated along the vertical direction of the flow, and the cells sedimentation was caused by the gravity force [47–50]. Capillary forces generated between the gas–liquid interfaces during the evaporation process were also employed as the power to let single cells drop into microwells [51, 52].

Meanwhile, specific materials were used to modify microchannels based on the differential adhesion effect with various cells. The patterned surface also attracted much attention due to its unique advantage for cells location [53–55]. Recently developed digital microfluidic devices [56–58], which integrated the substrate with electric circuit, were able to move, locate, and culture cells by manipulating liquid drops. Additionally, the methods of fabricating microfilter [59] and micropliers [60, 61], modificating antibodies [62], enzymes [63], or aptamers [64, 65] in microchannels inner surface were also developed to trap and manipulate cells. These approaches are high selective, easy to fabricate, and don't require special and expensive equipments, which enhanced the development and wide application.

1.3.2   Cell Culture on Microfluidic Devices

The *in vitro* cell culture in microfluidic devices require suitable microenvironment, sufficient nutrition and air, and the same humidity and temperature which mimic the *in vivo* conditions. According to the different research demands, two-dimensional and three-dimensional cell culture methods have been currently developed.

The most common method for cell culture is the two-dimensional culture. The substrate is coated with a layer of chemicals in order to achieve the adhesion of cells, such as poly-*L*-lysine [66] and collagen [67]. The patterned surface is usually constructed for a two cell co-culture, in order to simulate conditions of neighbored tissues in living organisms. The chemicals, which inhibit cell adhesion, could also be modified on the surface to pattern the cell growth region [9]. Based on two-dimensional culture models, specially designed structures were introduced to reduce the cell's shear-induced behavior during the culture medium injection [68].

The three-dimensional structures are fabricated in microchannels by the extra cellular matrix (ECM), in order to simulate the in vivo cell culture [54, 69, 70]. In that way the mechanical stress, elastic tension force, and fluid shear stress on the cells in the organs were simulated more realistically. And the cell-cell interactions and signal transmissions were better evaluated *via* three-dimensional approaches. Therefore, the three-dimensional cell culture was often used in cell co-culture and tissue engineering projects [7, 71]. Huh [72] and co-worker cultured the alveolar epithelial cells and endothelial cells on two sides of the porous membrane, which are modified by ECM, in order to simulate the breathing process of the alveolar. Nanoparticles were introduced to stimulate the cells, and the alveolar cells inflammation process was recorded and analyzed. Chung [8] et al. developed a three-dimensional network microfluidic device for cell co-culture. Different cell culturing regions were separated by the scaffolds made of ECM. The transmission of the signal factors in the vascular was simulated to study the cell migration effect during the co-culture process.

The laminar flow and diffusion theory in microfluidic devices were used to easily produce a continuous concentration gradient [73–75], which provided convenient conditions for exploring the cell culture microenvironment. Hung [76] and co-workers created the concentration gradient in a 10×10 culture chamber array, which achieved a long-term cell culture in a microfluidic device (Fig. 1.4). Culture medium was perfused from both the x and y axis direction to refresh the medium or subculture the cells. Different growth factors were added in the direction, which the concentration gradient generated, thus more elements during the inducing process for cells growth could be examined [77]. Based on the laminar flow effect, a mechanical hardness gradient was created by collagen gel on the substrate, in order to evaluate different factors on the regulation of the cells growth [78].

Various culture substrates and materials have been developed for the cell culturing on microfluidic devices. Huang [79] et al. fabricated a transparent microtube array for cell culture. Jung [80] and co-workers integrated polyurethane on both sides of a membrane in microchannels and cultured the cells attaching on both sides.

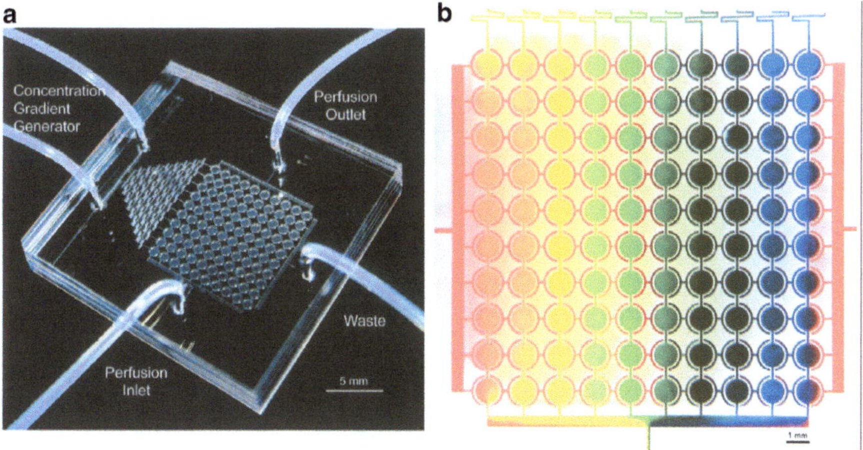

Fig. 1.4 A concentration gradient was created to control the microenvironment of a cell culture. (**a**) Layout of the microchip. (**b**) Concentration gradient generated in the microchannels (Reproduced from Ref. [76], with the permission of John Wiley and Sons Ltd.)

1.4   Cell Sorting and Separation on Microfluidic Devices

Cell sorting and separation are essential steps in cell biology research and in many diagnostic and therapeutic methods, because of the potential interference caused by the complex mixture of the matrix. The sample of interest normally contains a heterogeneous population of cells in culture or comprises a tissue, which affects the accurate analysis of a specific type of cells. Thus sorting and separation are necessary prior to the stimulation and detection in order to obtain accurate results.

Two of the commonly used techniques are fluorescence-activated cell sorting (FACS) [81, 82] and magnetic-activated cell sorting (MACS). FACS is an active sorting method, which utilizes complementary fluorophore-conjugated antibodies to label cells of interests. Collected scatter and fluorescence data are analyzed to identify cell types, numbers, and other parameters of interests. A high-throughput of $\sim$30,000 cells/s can be achieved by current commercial systems. MACS is a passive separation technique, which employs antibody-conjugated magnetic beads to bind specific proteins on the target cells. An external magnetic field is applied to drive the tagged cells and direct them to a collector [83].

Various label-free cell sorting and separation approaches of microfluidic devices were developed. Laminar flow, gravity, and perturbation flow caused by microstrucutres were employed to sort the cells based on their differences in size, density, and shape. The comparison of several label-free cell sorting and separation methods are listed in Table 1.1. Most of the techniques are based on the differences in cells sizes, which are normally applied in the separation of blood cells from the plasma, and the screening of circulating tumor cells (CTCs). These approaches don't require special biological reagents, and using simple fabricated microchips is a good

Table 1.1 Comparison of several label-free cell sorting and separation methods

Method	Mode of sorting	Sorting resolution	Throughput	Sorting efficiency	References
Mechanical filters	Size exclusion	17 ± 1.5 μm	0.75 mL/min	90%	[84]
Hydrodynamic	Streamline	–	10^5 cells/min	–	[85, 86]
Microstructures	Perturbation flow	<3 μm	4 μL/min	95.5%	[16]
Inertial	Lift forces and secondary flows	~5 μm	~10^6 cell/min	~80%	[87, 88]
Gravity	Sedimentation differences	~17 μm	~17 μL/min	99.97%	[89]
Biomimetic	Mimicking micro-vasculature	–	~0.2 μL/min	15–20% of plasma isolated	[90]
Magnetic	Differential magnetophoretic mobility	Intrinsic magnetic susceptibility	216–450 μL/min	95%	[91]
ATPs	Differential affinity	Surface properties and net charge	1 μL/min	–	[92]
Acoustic	Acoustic radiation force	>1 μm	80 μL/min	100%	[93]
Dielectrophoretic	Dielectophoretic force	~0.5 μm	<6 μL/min	>90%	[94]

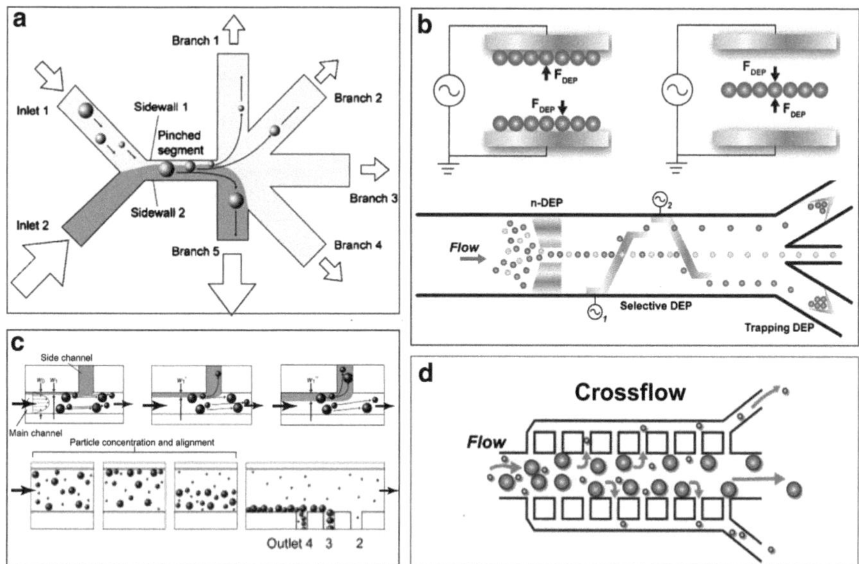

Fig. 1.5 Label-free cell sorting and separation in microfluidic devices: (a) Pinched flow fractionation separation (Reprinted from Ref. [95], with kind permission from Springer Science + Business Media). (b) Dielectrophoretic separation (Adapted with permission from Ref. [94]. Copyright 2008 American Chemical Society). (c) Hydrodynamic filtration (Adapted with permission from Ref. [96]. Copyright 2006 American Chemical Society). (d) Cross-flow filter (Reprinted from Ref. [97], with kind permission from Springer Science+Business Media)

choice. The fast separation speed and large analysis capacity are also important for a high-throughput sorting and separation, which can be achieved with microchips. However, extra marker is still required for specific selection of a particular target type of cells. Several label-free cell sorting methods [94–97] are shown in Fig. 1.5.

Specific small molecules or proteins on the cell membranes were used to identify cell species, in order to achieve the capture and separation of different cells. There are mainly three modes for cell sorting based on specific marker molecules: (1) Specific adsorption by protein identification: For example, the epithelial membrane antigen EMA and epidermal growth factor receptor EGFR could be treated as the tumor marker to identify breast cancer cells [98–100]. (2) Specific adsorption by peptides identification: Peptides with a specific sequence patterned on a microfluidic device surface were employed to catch the cells by specific bonding [101, 102]. (3) Specific adsorption by aptamers identification [103, 104]: The selected aptamers could specific recognize target cell surfaces [62, 105, 106]. Highly accurate identification could be achieved on various cells without knowing detailed information of the molecules on the cell membrane. This technique is widely applied in the identification and separation of human erythrocytes [107], cancer cells [108–110], and stem cells [111–113] (Fig. 1.6).

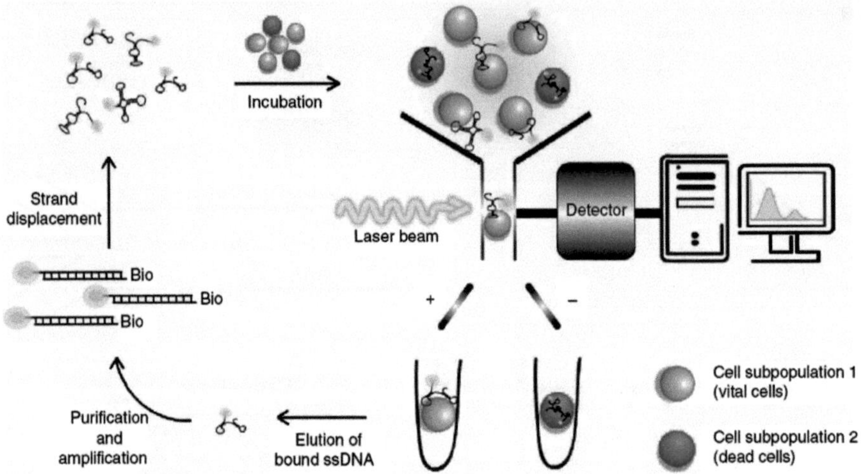

Fig. 1.6 Different cells were separated based on the specific identification of aptamers (Reprinted by permission from Macmillan Publishers Ltd: Ref. [111]. Copyright 2010)

1.5 Cell Stimulation and Detection on Microfluidic Devices

1.5.1 Cell Stimulation on Microfluidic Devices

After being trapped and cultured in microfluidic devices, cells are stimulated by the integrated function unit, in order to obtain a large amount of information of the cell response. Because of the wide selection range of the materials for microchips, and multiple design possibilities and various stimulations could be achieved in microfluidic devices. Fluid shear stress, mechanical stress, laser, temperature, and chemicals are able to be applied on the cells to obtain a stimulation reaction in certain designs.

Regarding different cell species different stimulation approaches were adopted. Since adherent cells attach the substrate tightly after being cultured, the solutions flowing over the cells can be replaced to achieve the stimulation. Commonly used methods are as follows: Different drug species and concentrations were applied to create various stimulations, and a concentration gradient was generated by the laminar flow effect, in order to structure a stimulation parameter array related to the location, which allowed for a rapid screening [114–117]. A shear stress was created by switching the flow speed, to observe the induction effect on cells [118, 119]. A temperature gradient was generated by laminar flow to evaluate cell response, as shown in Fig. 1.7b. Lucchetta [120] and co-workers located the embryos of fruit fly under the interface of a laminar flow with different temperatures, and various growth speeds were observed.

Since suspension cells would flow together within the culturing media, special designed structures are demanded to create a movement of the cells with the flow

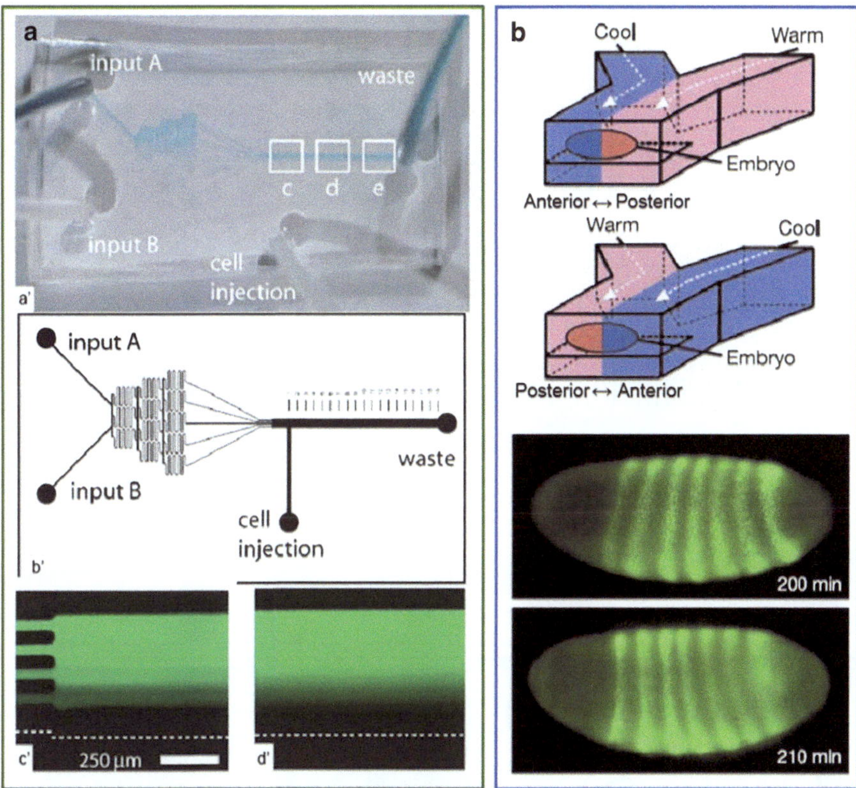

Fig. 1.7 Stimulation of adherent cells in microfluidic devices: (**a**) Concentration gradient induced stimulation generated by a laminar flow (Reprinted by permission from Macmillan Publishers Ltd: Ref. [114]. Copyright 2009). (**b**) Growth differences of embryos under differential temperature (Reprinted by permission from Macmillan Publishers Ltd: Ref. [120]. Copyright 2005)

[17, 121]. Yang [40] and co-workers fabricated a microdam in the channels, to temporarily fix suspension cells through the flow dynamic. Both, the concentration gradient and shear force were applied on HL-60 cells, in order to monitor the calcium ion release under ATP induction. Moreover, suspension cells were fixed by special designed structures, followed by desired manipulations. Gilleland [122] et al. integrated a membrane between microchannels in different layers in a microchip, in order to press and fix the nematode. Further operations by a laser to cut the nerve were carried out to observe the damage on its movement behavior (Fig. 1.8a).

1.5.2   Cell Secretion Detection on Microfluidic Devices

Different analysis objects require different detection approaches. A large amount of new techniques and new methods are currently developed for cell secretion

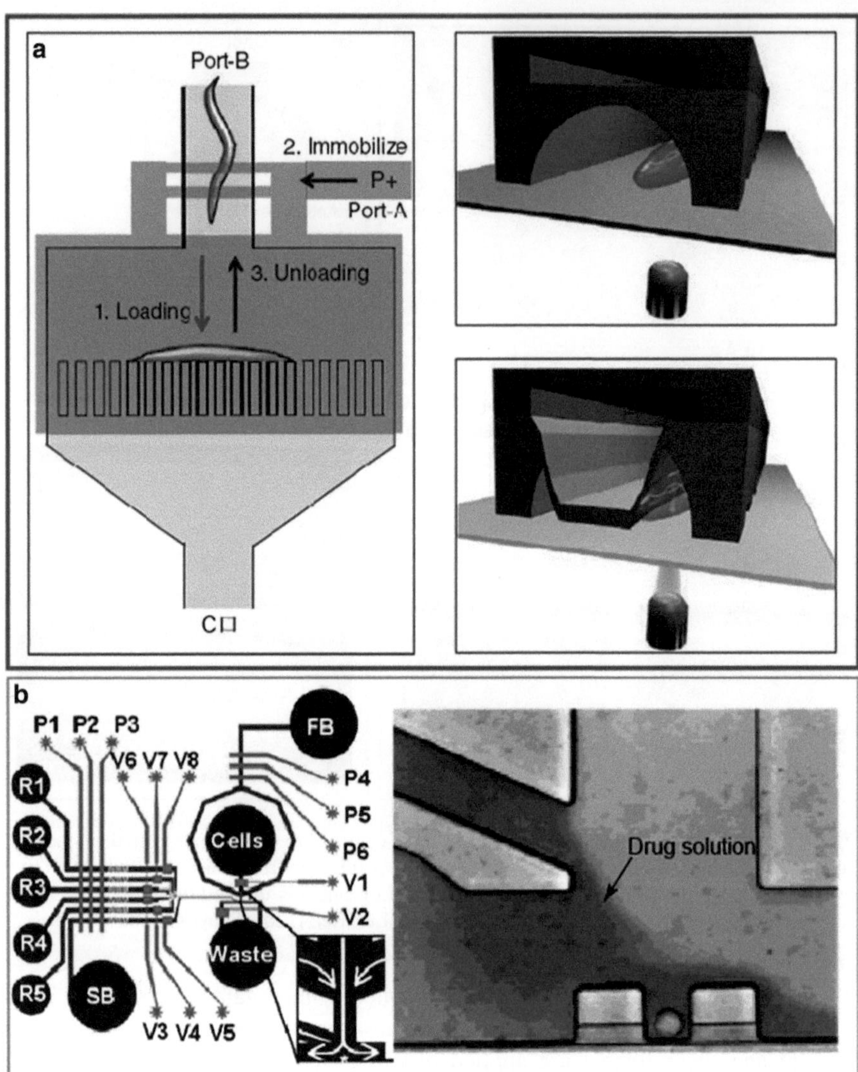

Fig. 1.8 Stimulation approaches on suspension cells on microfluidic devices: (**a**) A fixed nematode was stimulated by a laser (Reprinted by permission from Macmillan Publishers Ltd: Ref. [122]. Copyright 2010). (**b**) A single suspension cell was trapped and a drug solution was applied (Adapted with permission from Ref. [17]. Copyright 2003 American Chemical Society)

analysis. The most direct method for observation is labeling the target analytes by fluorescence markers, and apply quantitative analysis by imaging techniques [123]. This approach was mainly adopted for intracellular molecule detection. Small molecule probes were employed to penetrate the cell membrane, which can bind specific compounds and emit fluoresce [124]. Immunoassays were widely applied

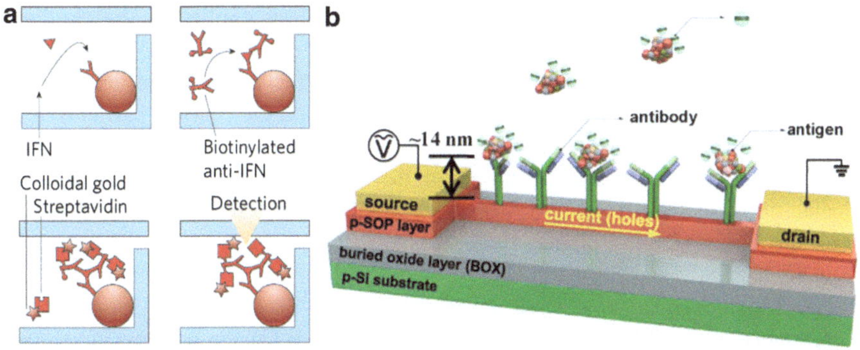

Fig. 1.9 Cell specific analysis based on the immunoassay: (**a**) Solid phase coated with antibodies was fixed in a microchannel to bind target compounds (Reproduced from Ref. [126], with the permission of John Wiley and Sons Ltd.) (**b**) FET-based sensor structure with immobilized antibodies and detection of negatively charged antigen (Reprinted from Ref. [127], with permission from Elsevier)

for protein analysis [125, 126] (Fig. 1.9a). Based on the immunoaffinity principle, a rapid enrichment of the target compounds was achieved even in complex matrix [128]. Furthermore a high throughput detection of cell secretion was carried out in the integrated microfluidic device [129]. The application of carbon nanotubes [130] and silicon nanowires [127, 131] as a sensor for target compounds further developed the miniaturization of the cell analysis in microfluidic devices (Fig. 1.9b).

To meet the upcoming requirement for higher sensitivity and precision in cell analysis, mass spectrometry (MS) started to play an more important role, especially in the qualitative analysis for small molecules and biological macromolecules. MS was able to provide the information of the molecules structure, and assistant for semi-quantitative analysis when it is required. The most commonly used MS ion sources for biological molecules analysis are electrospray ionization (ESI) and matrix assisted laser desorption ionization (MALDI), while the mainly employed mass analyzers are time-of-flight (TOF) or quadrupole time-of-flight (Q-TOF). A large number of achievements were obtained in the research of peptides and proteins using MS [132, 133].

The ESI source is suitable for the analysis of various peptides and proteins in biological samples, and the unique ionization is able to be integrated in microfluidic devices to achieve an on-line detection, thus it is widely used for biological analysis. In recent years, many research groups are focusing on the combination of microfluidic devices and MS by fabricating the interface for ESI source in microchips. The most important developed methods are as follows:

1. Combination of ESI source and capillary electrophoresis: The end of the capillary of the microfluidic device was formed as a sharp needle, to replace the traditional needle producing the spray in the ion source of the ESI. Bings and co-worker [134] first published the approach to directly connect a capillary needle for

nebulization at the end of the glass chip, which was applied to the capillary electrophoresis separation. The accurate detection results required a very small dead volume, which could be easily realized on the microfluidic device. The bonding efficiency between the target molecules and the charged particles after nebulization exerted a significant impact on the detection results, as well as the bandwidth of each compound in the mixture after being separated by electrophoresis. This method was widely applied in the separation and analysis of medicines [135], proteins [136], and the composition of metabolic processes [137, 138].

2. On-chip ionization in microfluidic devices: The electron spray was directly produced at the exit of the microchannels. At first, the technique was developed to create the electron spray microchannel exit, which was located at the edge of the glass chip [139, 140]. However, the hydrophilic glass surface could easily cause moistening of the edge of the microchannel exit, greatly increase the dead volume, and reduce the separation efficiency. Xue [139] and co-workers made efforts on the hydrophobic treatment of the microchannel exit to overcome this problem. Several materials such as PDMS [141] and PET [142] were also modified on the microchannel exit. The modification increased the complexity of fabrication process, but also shortened the lifetime of the microchip.

3. External spray interface on microfluidic devices (Fig. 1.10): An external spray interface was fabricated on a microchip to achieve the ionization [142, 147–149]. Drilling followed by gluing was commonly applied to combine the spray interface and microchip. Compared to the on-chip spray interface, it was easier to obtain a high voltage and a small dead volume for the external spray interface. The composition in the desired location of the microchannel, such as the microchannel outlet [150], flow junction [151], and sheath flow channel [152], was able to be detected through direct nebulization. The simple fabrication, low reagent consumption, and flexible design results in the currently wide application of this approach.

On-line real-time monitoring can be achieved by connecting the ESI source and the microfluidic devices, while the highly accurate analysis can be carried out by a TOF mass analyzer. However, due to the small amount of the target compounds, a great impact will be caused on the precision and accuracy of the results only by a trace of impurities. Thus, sample purification is required before the MS detection. Salt ions and high-abundance interference proteins need to be removed, to prevent interferences on the accurate qualitative and quantitative analysis of the target compounds. Many works have been published to integrate a pretreatment unit for biological samples in microfluidic devices [153, 154]. Monolithic columns were integrated in microchannels through the light-initiated polymerization, to realize the on-line sample purification prior to detection [155]. Commercial solid-phase extraction packing materials were integrated in microchips, in order to desalinate and enrich cell secretions [137].

The MALDI source is more applicable for biological macromolecules, such as protein analysis. The molecular ion peak could be obtained, and the protein

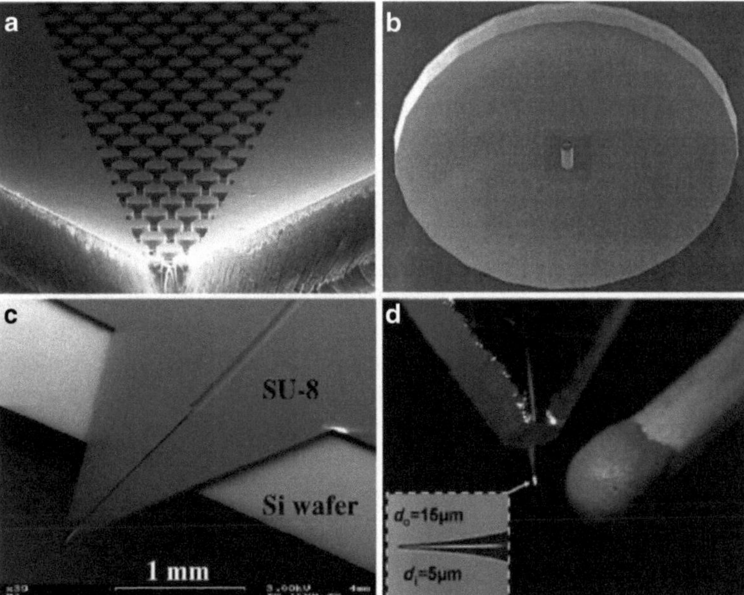

Fig. 1.10 External spray interfaces on microfluidic devices: (**a**) Both sides were etched to sharp the interface (Reprinted from Ref. [143], with permission from Elsevier). (**b**) A silicon nozzle on the flat microchip surface (Reprinted with permission from Ref. [144]. Copyright 2003 American Chemical Society). (**c**) External SU-8 needle (Reprinted from Ref. [145], with permission from Elsevier). (**d**) Glass needle integrated in the microchip (Reprinted from Ref. [146], with the permission of John Wiley and Sons Ltd.)

structure could be analyzed by enzymolysis. But it's difficult to develop MALDI as an on-line detection approach, and high sample purity is required, which leads to a more complicated purification process compared to ESI. Jo and co-workers [156] modified C18 particles on the silicon surface, in order to purify and concentrate the neuropeptide released by neuronal cells. MALDI detection was applied after eluting the peptide, and the neuronal cells respond to the stimulation with potassium chloride solution was evaluated.

1.6   Research Orientation and Significance of This Thesis

Microfluidic devices have tremendous potential for a wide range of applications in the future, particularly in the early diagnosis and medical screening. Especially in less developed countries, the early signs of certain diseases are ignored, because of the lack of available techniques. Infectious diseases cannot be diagnosed in an early state, which can cause serious damages, because of a late medical treatment. The developed techniques based on microfluidic devices, which are portable, have a low power consumption and are simple to handle, are promoted and propagated in medical care in less developed regions. It is also an effective method to control global epidemic. Moreover, for the point-of-care diagnosis and family health care, the development of portable medical equipments and rapid diagnosis approaches still require further development [157].

This thesis focused on the cell metabolism analysis based on microfluidic devices. A new analysis platform combining microfluidic devices with high sensitive MS was established. First, a single particle was used as a model of the cell, in order to build up the combined analysis platform, and achieve the single particle capture and fixation on a microfluidic chip. The contents of herbicides in environmental samples were measured, while the accuracy and reproducibility of the analysis platform were evaluated. Based on the established platform, cancer cells were fixed, cultured, and stimulated in the microfluidic chips. The micro-columns for extraction were integrated into the microfluidic system, in order to purify and enrich the cell secretion prior to the MS detection. Medical effects were evaluated by monitoring the contents of cell secretion.

After the establishment of the analysis platform combined with microfluidic devices and MS, a new problem has arisen: The matrix of different cells is too complex to be analyzed with MS. To deal with complex real biological samples, microfluidic devices for cell separation were developed to solve this problem. The target cells were filtered for the following fixation and detection. Thereby the integrated, automatic, and high throughput cell analysis on microfluidic devices was further promoted. In order to better explain the responds of living organs under treatment with drugs, the complex *in vivo* micro-environment around the cells was simulated, and the interactions and signaling mechanisms between cells were further studied. This thesis focused on the establishment of the analysis platform, combining microfluidic devices and MS, the cells fixation in microchannels, and the purification and enrichment of cell secretions. The cell sorting device was developed to classify complex real samples, and the cell signaling transmission was studied to obtain more useful information to solve biological and medical problems.

References

1. Manz A, Graber N, Widmer HM (1990) Miniaturized total chemical-analysis systems – a novel concept for chemical sensing. Sens Actuat B Chem 1:244–248
2. Emrich CA, Tian HJ, Medintz IL, Mathies RA (2002) Microfabricated 384-lane capillary array electrophoresis bioanalyzer for ultrahigh-throughput genetic analysis. Anal Chem 74:5076–5083
3. Thorsen T, Maerkl SJ, Quake SR (2002) Microfluidic large-scale integration. Science 298:580–584
4. Vilkner T, Janasek D, Manz A (2004) Micro total analysis systems. Recent developments. Anal Chem 76:3373–3385
5. Liang SL, Chan DW (2007) Enzymes and related proteins as cancer biomarkers: a proteomic approach. Clin Chim Acta 381:93–97

6. Huang CP, Lu J, Seon H, Lee AP, Flanagan LA, Kim HY, Putnam AJ, Jeon NL (2009) Engineering microscale cellular niches for three-dimensional multicellular co-cultures. Lab Chip 9:1740–1748

7. Sung JH, Shuler ML (2009) A micro cell culture analog (mu CCA) with 3-D hydrogel culture of multiple cell lines to assess metabolism-dependent cytotoxicity of anti-cancer drugs. Lab Chip 9:1385–1394

8. Chung S, Sudo R, Mack PJ, Wan CR, Vickerman V, Kamm RD (2009) Cell migration into scaffolds under co-culture conditions in a microfluidic platform. Lab Chip 9:269–275

9. Tourovskaia A, Figueroa-Masot X, Folch A (2005) Differentiation-on-a-chip: a microfluidic platform for long-term cell culture studies. Lab Chip 5:14–19

10. Yamada M, Hirano T, Yasuda M, Seki M (2006) A microfluidic flow distributor generating stepwise concentrations for high-throughput biochemical processing. Lab Chip 6:179–184

11. Paguirigan A, Beebe DJ (2006) Gelatin based microfluidic devices for cell culture. Lab Chip 6:407–413

12. Toh YC, Zhang C, Zhang J, Khong YM, Chang S, Samper VD, van Noort D, Hutmacher DW, Yu HR (2007) A novel 3D mammalian cell perfusion-culture system in microfluidic channels. Lab Chip 7:302–309

13. Gomez-Sjoberg R, Leyrat AA, Pirone DM, Chen CS, Quake SR (2007) Versatile, fully automated, microfluidic cell culture system. Anal Chem 79:8557–8563

14. Yao B, Luo GA, Feng X, Wang W, Chen LX, Wang YM (2004) A microfluidic device based on gravity and electric force driving for flow cytometry and fluorescence activated cell sorting. Lab Chip 4:603–607

15. Pamme N, Wilhelm C (2006) Continuous sorting of magnetic cells via on-chip free-flow magnetophoresis. Lab Chip 6:974–980

16. Choi S, Song S, Choi C, Park JK (2009) Microfluidic self-sorting of mammalian cells to achieve cell cycle synchrony by hydrophoresis. Anal Chem 81:1964–1968

17. Wheeler AR, Throndset WR, Whelan RJ, Leach AM, Zare RN, Liao YH, Farrell K, Manger ID, Daridon A (2003) Microfluidic device for single-cell analysis. Anal Chem 75:3581–3586

18. Bailey RC, Kwong GA, Radu CG, Witte ON, Heath JR (2007) DNA-encoded antibody libraries: a unified platform for multiplexed cell sorting and detection of genes and proteins. J Am Chem Soc 129:1959–1967

19. McClain MA, Culbertson CT, Jacobson SC, Allbritton NL, Sims CE, Ramsey JM (2003) Microfluidic devices for the high-throughput chemical analysis of cells. Anal Chem 75:5646–5655

20. Lee JN, Jiang X, Ryan D, Whitesides GM (2004) Compatibility of mammalian cells on surfaces of poly(dimethylsiloxane). Langmuir 20:11684–11691

21. Piruska A, Nikcevic I, Lee SH, Ahn C, Heineman WR, Limbach PA, Seliskar CJ (2005) The autofluorescence of plastic materials and chips measured under laser irradiation. Lab Chip 5:1348–1354

22. Wu HK, Ren KN, Zhao YH, Su J, Ryan D (2010) Convenient method for modifying poly(dimethylsiloxane) to be airtight and resistive against absorption of small molecules. Anal Chem 82:5965–5971

23. Lai S, Wang SN, Luo J, Lee LJ, Yang ST, Madou MJ (2004) Design of a compact disk-like microfluidic platform for enzyme-linked immunosorbent assay. Anal Chem 76:1832–1837

24. Mehta G, Lee J, Cha W, Tung YC, Linderman JJ, Takayama S (2009) Hard Top soft bottom microfluidic devices for cell culture and chemical analysis. Anal Chem 81:3714–3722

25. Bettinger CJ, Weinberg EJ, Kulig KM, Vacanti JP, Wang YD, Borenstein JT, Langer R (2006) Three-dimensional microfluidic tissue-engineering scaffolds using a flexible biodegradable polymer. Adv Mater 18:165–169

26. Fidkowski C, Kaazempur-Mofrad MR, Borenstein J, Vacanti JP, Langer R, Wang YD (2005) Endothelialized microvasculature based on a biodegradable elastomer. Tissue Eng 11:302–309

27. Heo J, Thomas KJ, Seong GH, Crooks RM (2003) A microfluidic bioreactor based on hydrogel-entrapped E. coli: cell viability, lysis, and intracellular enzyme reactions. Anal Chem 75:22–26

28. Ling Y, Rubin J, Deng Y, Huang C, Demirci U, Karp JM, Khademhosseini A (2007) A cell-laden microfluidic hydrogel. Lab Chip 7:756–762
29. Liu MC, Ho D, Tai YC (2008) Monolithic fabrication of three-dimensional microfluidic networks for constructing cell culture array with an integrated combinatorial mixer. Sens Actuat B Chem 129:826–833
30. Kim MJ, Breuer KS (2008) Microfluidic pump powered by self-organizing bacteria. Small 4:111–120
31. Meyvantsson I, Warrick JW, Hayes S, Skoien A, Beebe DJ (2008) Automated cell culture in high density tubeless microfluidic device arrays. Lab Chip 8:717–724
32. Unger MA, Chou HP, Thorsen T, Scherer A, Quake SR (2000) Monolithic microfabricated valves and pumps by multilayer soft lithography. Science 288:113–116
33. Hulme SE, Shevkoplyas SS, Whitesides GM (2009) Incorporation of prefabricated screw, pneumatic, and solenoid valves into microfluidic devices. Lab Chip 9:79–86
34. King KR, Wang SH, Irimia D, Jayaraman A, Toner M, Yarmush ML (2007) A high-throughput microfluidic real-time gene expression living cell array. Lab Chip 7:77–85
35. Berger M, Castelino J, Huang R, Shah M, Austin RH (2001) Design of a microfabricated magnetic cell separator. Electrophoresis 22:3883–3892
36. Lee H, Purdon AM, Westervelt RM (2004) Manipulation of biological cells using a microelectromagnet matrix. Appl Phys Lett 85:1063–1065
37. Furdui VI, Harrison DJ (2004) Immunomagnetic T cell capture from blood for PCR analysis using microfluidic systems. Lab Chip 4:614–618
38. Umehara S, Wakamoto Y, Inoue I, Yasuda K (2003) On-chip single-cell microcultivation assay for monitoring environmental effects on isolated cells. Biochem Biophys Res Commun 305:534–540
39. Enger J, Goksor M, Ramser K, Hagberg P, Hanstorp D (2004) Optical tweezers applied to a microfluidic system. Lab Chip 4:196–200
40. Yang MS, Li CW, Yang J (2002) Cell docking and on-chip monitoring of cellular reactions with a controlled concentration gradient on a microfluidic device. Anal Chem 74:3991–4001
41. Valero A, Merino F, Wolbers F, Luttge R, Vermes I, Andersson H, van den Berg A (2005) Apoptotic cell death dynamics of HL60 cells studied using a microfluidic cell trap device. Lab Chip 5:49–55
42. Kobel S, Valero A, Latt J, Renaud P, Lutolf M (2010) Optimization of microfluidic single cell trapping for long-term on-chip culture. Lab Chip 10:857–863
43. Ogunniyi AO, Story CM, Papa E, Guillen E, Love JC (2009) Screening individual hybridomas by microengraving to discover monoclonal antibodies. Nat Protoc 4:767–782
44. Di Carlo D, Wu LY, Lee LP (2006) Dynamic single cell culture array. Lab Chip 6:1445–1449
45. Skelley AM, Kirak O, Suh H, Jaenisch R, Voldman J (2009) Microfluidic control of cell pairing and fusion. Nat Methods 6:147–152
46. Frimat JP, Becker M, Chiang YY, Marggraf U, Janasek D, Hengstler JG, Franzke J, West J (2011) A microfluidic array with cellular valving for single cell co-culture. Lab Chip 11:231–237
47. Liu CS, Liu JJ, Gao D, Ding MY, Lin JM (2010) Fabrication of microwell arrays based on two-dimensional ordered polystyrene microspheres for high-throughput single-cell analysis. Anal Chem 82:9418–9424
48. Lee SH, Jeong HE, Park MC, Hur JY, Cho HS, Park SH, Suh KY (2008) Fabrication of hollow polymeric microstructures for shear-protecting cell containers. Adv Mater 20:788–792
49. Khademhosseini A, Yeh J, Eng G, Karp J, Kaji H, Borenstein J, Farokhzad OC, Langer R (2005) Cell docking inside microwells within reversibly sealed microfluidic channels for fabricating multiphenotype cell arrays. Lab Chip 5:1380–1386
50. Rettig JR, Folch A (2005) Large-scale single-cell trapping and imaging using microwell arrays. Anal Chem 77:5628–5634
51. Park MC, Hur JY, Cho HS, Park SH, Suh KY (2011) High-throughput single-cell quantifi-cation using simple microwell-based cell docking and programmable time-course live-cell imaging. Lab Chip 11:79–86

52. Park MC, Hur JY, Kwon KW, Park SH, Suh KY (2006) Pumpless, selective docking of yeast cells inside a microfluidic channel induced by receding meniscus. Lab Chip 6:988–994
53. Anselme K, Davidson P, Popa AM, Giazzon M, Liley M, Ploux L (2010) The interaction of cells and bacteria with surfaces structured at the nanometre scale. Acta Biomater 6:3824–3846
54. Chen CS, Mrksich M, Huang S, Whitesides GM, Ingber DE (1997) Geometric control of cell life and death. Science 276:1425–1428
55. Kim P, Kim DH, Kim B, Choi SK, Lee SH, Khademhosseini A, Langer R, Suh KY (2005) Fabrication of nanostructures of polyethylene glycol for applications to protein adsorption and cell adhesion. Nanotechnology 16:2420–2426
56. Barbulovic-Nad I, Au SH, Wheeler AR (2010) A microfluidic platform for complete mammalian cell culture. Lab Chip 10:1536–1542
57. Abdelgawad M, Wheeler AR (2009) The digital revolution: a new paradigm for microfluidics. Adv Mater 21:920–925
58. Au SH, Shih SCC, Wheeler AR (2011) Integrated microbioreactor for culture and analysis of bacteria, algae and yeast. Biomed Microdevices 13:41–50
59. Yang J, Li CW, Yang MS (2004) Hydrodynamic simulation of cell docking in microfluidic channels with different dam structures. Lab Chip 4:53–59
60. Huang LR, Cox EC, Austin RH, Sturm JC (2004) Continuous particle separation through deterministic lateral displacement. Science 304:987–990
61. Chronis N, Lee LP (2005) Electrothermally activated SU-8 microgripper for single cell manipulation in solution. J Microelectromech Syst 14:857–863
62. Revzin A, Sekine K, Sin A, Tompkins RG, Toner M (2005) Development of a microfabricated cytometry platform for characterization and sorting of individual leukocytes. Lab Chip 5:30–37
63. Murthy SK, Sin A, Tompkins RG, Toner M (2004) Effect of flow and surface conditions on human lymphocyte isolation using microfluidic chambers. Langmuir 20:11649–11655
64. Xu Y, Phillips JA, Yan JL, Li QG, Fan ZH, Tan WH (2009) Aptamer-based microfluidic device for enrichment, sorting, and detection of multiple cancer cells. Anal Chem 81:7436–7442
65. Phillips JA, Xu Y, Xia Z, Fan ZH, Tan WH (2009) Enrichment of cancer cells using aptamers immobilized on a microfluidic channel. Anal Chem 81:1033–1039
66. Wei HB, Li HF, Gao D, Lin JM (2010) Multi-channel microfluidic devices combined with electrospray ionization quadrupole time-of-flight mass spectrometry applied to the monitoring of glutamate release from neuronal cells. Analyst 135:2043–2050
67. Camelliti P, McCulloch AD, Kohl P (2005) Microstructured cocultures of cardiac myocytes and fibroblasts: a two-dimensional in vitro model of cardiac tissue. Microsc Microanal 11:249–259
68. Park ES, Brown AC, DiFeo MA, Barker TH, Lu H (2010) Continuously perfused, non-cross-contaminating microfluidic chamber array for studying cellular responses to orthogonal combinations of matrix and soluble signals. Lab Chip 10:571–580
69. Tsang VL, Bhatia SN (2004) Three-dimensional tissue fabrication. Adv Drug Deliv Rev 56:1635–1647
70. Tan JL, Tien J, Pirone DM, Gray DS, Bhadriraju K, Chen CS (2003) Cells lying on a bed of microneedles: an approach to isolate mechanical force. Proc Natl Acad Sci USA 100:1484–1489
71. Kunze A, Giugliano M, Valero A, Renaud P (2011) Micropatterning neural cell cultures in 3D with a multi-layered scaffold. Biomaterials 32:2088–2098
72. Huh D, Matthews BD, Mammoto A, Montoya-Zavala M, Hsin HY, Ingber DE (2010) Reconstituting organ-level lung functions on a chip. Science 328:1662–1668
73. Walsh CL, Babin BM, Kasinskas RW, Foster JA, McGarry MJ, Forbes NS (2009) A multipurpose microfluidic device designed to mimic microenvironment gradients and develop targeted cancer therapeutics. Lab Chip 9:545–554
74. Cheng SY, Heilman S, Wasserman M, Archer S, Shuler ML, Wu MM (2007) A hydrogel-based microfluidic device for the studies of directed cell migration. Lab Chip 7:763–769

75. Glawdel T, Elbuken C, Lee LEJ, Ren CL (2009) Microfluidic system with integrated electroosmotic pumps, concentration gradient generator and fish cell line (RTgill-W1)-towards water toxicity testing. Lab Chip 9:3243–3250
76. Hung PJ, Lee PJ, Sabounchi P, Lin R, Lee LP (2005) Continuous perfusion microfluidic cell culture array for high-throughput cell-based assays. Biotechnol Bioeng 89:1–8
77. Chung BG, Flanagan LA, Rhee SW, Schwartz PH, Lee AP, Monuki ES, Jeon NL (2005) Human neural stem cell growth and differentiation in a gradient-generating microfluidic device. Lab Chip 5:401–406
78. Sundararaghavan HG, Monteiro GA, Firestein BL, Shreiber DI (2009) Neurite growth in 3D collagen gels with gradients of mechanical properties. Biotechnol Bioeng 102:632–643
79. Huang GS, Mei YF, Thurmer DJ, Coric E, Schmidt OG (2009) Rolled-up transparent microtubes as two-dimensionally confined culture scaffolds of individual yeast cells. Lab Chip 9:263–268
80. Jung JH, Choi CH, Chung S, Chung YM, Lee CS (2009) Microfluidic synthesis of a cell adhesive Janus polyurethane microfiber. Lab Chip 9:2596–2602
81. Fu AY, Chou HP, Spence C, Arnold FH, Quake SR (2002) An integrated microfabricated cell sorter. Anal Chem 74:2451–2457
82. Wang MM, Tu E, Raymond DE, Yang JM, Zhang HC, Hagen N, Dees B, Mercer EM, Forster AH, Kariv I, Marchand PJ, Butler WF (2005) Microfluidic sorting of mammalian cells by optical force switching. Nat Biotechnol 23:83–87
83. Liu YJ, Guo SS, Zhang ZL, Huang WH, Baigl D, Xie M, Chen Y, Pang DW (2007) A micropillar-integrated smart microfluidic device for specific capture and sorting of cells. Electrophoresis 28:4713–4722
84. Zheng S, Lin H, Liu JQ, Balic M, Datar R, Cote RJ, Tai YC (2007) Membrane microfilter device for selective capture, electrolysis and genomic analysis of human circulating tumor cells. J Chromatogr A 1162:154–161
85. Andersen KB, Levinsen S, Svendsen WE, Okkels F (2009) A generalized theoretical model for "continuous particle separation in a microchannel having asymmetrically arranged multiple branches". Lab Chip 9:1638–1639
86. Yamada M, Kano K, Tsuda Y, Kobayashi J, Yamato M, Seki M, Okano T (2007) Microfluidic devices for size-dependent separation of liver cells. Biomed Microdevices 9:637–645
87. Kuntaegowdanahalli SS, Bhagat AAS, Kumar G, Papautsky I (2009) Inertial microfluidics for continuous particle separation in spiral microchannels. Lab Chip 9:2973–2980
88. Wu ZG, Willing B, Bjerketorp J, Jansson JK, Hjort K (2009) Soft inertial microfluidics for high throughput separation of bacteria from human blood cells. Lab Chip 9:1193–1199
89. Huh D, Bahng JH, Ling YB, Wei HH, Kripfgans OD, Fowlkes JB, Grotberg JB, Takayama S (2007) Gravity-driven microfluidic particle sorting device with hydrodynamic separation amplification. Anal Chem 79:1369–1376
90. Shevkoplyas SS, Yoshida T, Munn LL, Bitensky MW (2005) Biomimetic autoseparation of leukocytes from whole blood in a microfluidic device. Anal Chem 77:933–937
91. Huang R, Barber TA, Schmidt MA, Tompkins RG, Toner M, Bianchi DW, Kapur R, Flejter WL (2008) A microfluidics approach for the isolation of nucleated red blood cells (NRBCs) from the peripheral blood of pregnant women. Prenat Diagn 28:892–899
92. SooHoo JR, Walker GM (2009) Microfluidic aqueous two phase system for leukocyte concentration from whole blood. Biomed Microdevices 11:323–329
93. Lenshof A, Ahmad-Tajudin A, Jaras K, Sward-Nilsson AM, Aberg L, Marko-Varga G, Malm J, Lilja H, Laurell T (2009) Acoustic whole blood plasmapheresis chip for prostate specific antigen microarray diagnostics. Anal Chem 81:6030–6037
94. Vahey MD, Voldman J (2008) An equilibrium method for continuous-flow cell sorting using dielectrophoresis. Anal Chem 80:3135–3143
95. Gossett DR, Weaver WM, Mach AJ, Hur SC, Kwong Tse HT, Lee W, Amini H, Di Carlo D (2010) Label-free cell separation and sorting in microfluidic systems. Anal Bioanal Chem 397:3249–3267

96. Yamada M, Seki M (2006) Microfluidic particle sorter employing flow splitting and recombining. Anal Chem 78:1357–1362

97. Ji HM, Samper V, Chen Y, Heng CK, Lim TM, Yobas L (2008) Silicon-based microfilters for whole blood cell separation. Biomed Microdevices 10:251–257

98. Evron E, Dooley WC, Umbricht CB, Rosenthal D, Sacchi N, Gabrielson E, Soito AB, Hung DT, Ljung BM, Davidson NE, Sukumar S (2001) Detection of breast cancer cells in ductal lavage fluid by methylation-specific PCR. Lancet 357:1335–1336

99. Fuqua SAW, Wiltschke C, Zhang QX, Borg A, Castles CG, Friedrichs WE, Hopp T, Hilsenbeck S, Mohsin S, O'Connell P, Allred DC (2000) A hypersensitive estrogen receptor-alpha mutation in premalignant breast lesions. Cancer Res 60:4026–4029

100. Du Z, Cheng KH, Vaughn MW, Collie NL, Gollahon LS (2007) Recognition and capture of breast cancer cells using an antibody-based platform in a microelectromechanical systems device. Biomed Microdevices 9:35–42

101. Plouffe BD, Njoka DN, Harris J, Liao JH, Horick NK, Radisic M, Murthy SK (2007) Peptide-mediated selective adhesion of smooth muscle and endothelial cells in microfluidic shear flow. Langmuir 23:5050–5055

102. Hasenbein ME, Andersen TT, Bizios R (2002) Micropatterned surfaces modified with select peptides promote exclusive interactions with osteoblasts. Biomaterials 23:3937–3942

103. Tuerk C, Gold L (1990) Systematic evolution of ligands by exponential enrichment – RNA ligands to bacteriophage-T4 DNA-polymerase. Science 249:505–510

104. Ellington AD, Szostak JW (1990) Invitro Selection of RNA Molecules That Bind Specific Ligands. Nature 346:818–822

105. Neryl AA, Wrenger C, Ulrich H (2009) Recognition of biomarkers and cell-specific molecular signatures: aptamers as capture agents. J Sep Sci 32:1523–1530

106. Chang WC, Lee LP, Liepmann D (2005) Biomimetic technique for adhesion-based collection and separation of cells in a microfluidic channel. Lab Chip 5:64–73

107. Morris KN, Jensen KB, Julin CM, Weil M, Gold L (1998) High affinity ligands from in vitro selection: complex targets. Proc Natl Acad Sci USA 95:2902–2907

108. Lien KY, Chuang YH, Hung LY, Hsu KF, Lai WW, Ho CL, Chou CY, Lee GB (2010) Rapid isolation and detection of cancer cells by utilizing integrated microfluidic systems. Lab Chip 10:2875–2886

109. Herr JK, Smith JE, Medley CD, Shangguan DH, Tan WH (2006) Aptamer-conjugated nanoparticles for selective collection and detection of cancer cells. Anal Chem 78:2918–2924

110. Shangguan D, Cao ZH, Meng L, Mallikaratchy P, Sefah K, Wang H, Li Y, Tan WH (2008) Cell-specific aptamer probes for membrane protein elucidation in cancer cells. J Proteome Res 7:2133–2139

111. Mayer G, Ahmed MSL, Dolf A, Endl E, Knolle PA, Famulok M (2010) Fluorescence-activated cell sorting for aptamer SELEX with cell mixtures. Nat Protoc 5:1993–2004

112. Guo KT, Schafer R, Paul A, Gerber A, Ziemer G, Wendel HP (2006) A new technique for the isolation and surface immobilization of mesenchymal stem cells from whole bone marrow using high-specific DNA aptamers. Stem Cells 24:2220–2231

113. Cerchia L, de Franciscis V (2010) Targeting cancer cells with nucleic acid aptamers. Trends Biotechnol 28:517–525

114. Jeon NL, Baskaran H, Dertinger SKW, Whitesides GM, Van de Water L, Toner M (2002) Neutrophil chemotaxis in linear and complex gradients of interleukin-8 formed in a microfabricated device. Nat Biotechnol 20:826–830

115. Takayama S, Ostuni E, LeDuc P, Naruse K, Ingber DE, Whitesides GM (2001) Laminar flows – Subcellular positioning of small molecules. Nature 411:1016–1016

116. Kim M, Kim T (2010) Diffusion-based and long-range concentration gradients of multiple chemicals for bacterial chemotaxis assays. Anal Chem 82:9401–9409

117. Walker GM, Sai JQ, Richmond A, Stremler M, Chung CY, Wikswo JP (2005) Effects of flow and diffusion on chemotaxis studies in a microfabricated gradient generator. Lab Chip 5:611–618

118. Powers MJ, Domansky K, Kaazempur-Mofrad MR, Kalezi A, Capitano A, Upadhyaya A, Kurzawski P, Wack KE, Stolz DB, Kamm R, Griffith LG (2002) A microfabricated array bioreactor for perfused 3D liver culture. Biotechnol Bioeng 78:257–269
119. Leclerc E, David B, Griscom L, Lepioufle B, Fujii T, Layrolle P, Legallaisa C (2006) Study of osteoblastic cells in a microfluidic environment. Biomaterials 27:586–595
120. Lucchetta EM, Lee JH, Fu LA, Patel NH, Ismagilov RF (2005) Dynamics of Drosophila embryonic patterning network perturbed in space and time using microfluidics. Nature 434:1134–1138
121. El-Ali J, Gaudet S, Gunther A, Sorger PK, Jensen KF (2005) Cell stimulus and lysis in a microfluidic device with segmented gas–liquid flow. Anal Chem 77:3629–3636
122. Gilleland CL, Rohde CB, Zeng F, Yanik MF (2010) Microfluidic immobilization of physiologically active Caenorhabditis elegans. Nat Protoc 5:1888–1902
123. Easley CJ, Rocheleau JV, Head WS, Piston DW (2009) Quantitative measurement of zinc secretion from pancreatic islets with high temporal resolution using droplet-based microfluidics. Anal Chem 81:9086–9095
124. Gao D, Liu JJ, Wei HB, Li HF, Guo GS, Lin JM (2010) A microfluidic approach for anticancer drug analysis based on hydrogel encapsulated tumor cells. Anal Chim Acta 665:7–14
125. Lion N, Rohner TC, Dayon L, Arnaud IL, Damoc E, Youhnovski N, Wu ZY, Roussel C, Josserand J, Jensen H, Rossier JS, Przybylski M, Girault HH (2003) Microfluidic systems in proteomics. Electrophoresis 24:3533–3562
126. Sato K, Yamanaka M, Takahashi H, Tokeshi M, Kimura H, Kitamori T (2002) Microchip-based immunoassay system with branching multichannels for simultaneous determination of interferon-gamma. Electrophoresis 23:734–739
127. Kim A, Ah CS, Park CW, Yang J-H, Kim T, Ahn C-G, Park SH, Sung GY (2010) Direct label-free electrical immunodetection in human serum using a flow-through-apparatus approach with integrated field-effect transistors. Biosens Bioelectron 25:1767–1773
128. Chen C, Skog J, Hsu CH, Lessard RT, Balaj L, Wurdinger T, Carter BS, Breakefield XO, Toner M, Irimia D (2010) Microfluidic isolation and transcriptome analysis of serum microvesicles. Lab Chip 10:505–511
129. Dishinger JF, Reid KR, Kennedy RT (2009) Quantitative monitoring of insulin secretion from single islets of Langerhans in parallel on a microfluidic chip. Anal Chem 81:3119–3127
130. Gruner G (2006) Carbon nanotube transistors for biosensing applications. Anal Bioanal Chem 384:322–335
131. Zheng GF, Patolsky F, Cui Y, Wang WU, Lieber CM (2005) Multiplexed electrical detection of cancer markers with nanowire sensor arrays. Nat Biotechnol 23:1294–1301
132. Gustafsson M, Hirschberg D, Palmberg C, Jornvall H, Bergman T (2004) Integrated sample preparation and MALDI mass spectrometry on a microfluidic compact disk. Anal Chem 76:345–350
133. Liu HH, Felten C, Xue QF, Zhang BL, Jedrzejewski P, Karger BL, Foret F (2000) Development of multichannel devices with an array of electrospray tips far high-throughput mass spectrometry. Anal Chem 72:3303–3310
134. Bings NH, Wang C, Skinner CD, Colyer CL, Thibault P, Harrison DJ (1999) Microfluidic devises connected to fused-silica capillaries with minimal dead volume. Anal Chem 71:3292–3296
135. Lazar IM, Grym J, Foret F (2006) Microfabricated devices: a new sample introduction approach to mass spectrometry. Mass Spectrom Rev 25:573–594
136. Wang C, Oleschuk R, Ouchen F, Li JJ, Thibault P, Harrison DJ (2000) Integration of immobilized trypsin bead beds for protein digestion within a microfluidic chip incorporating capillary electrophoresis separations and an electrospray mass spectrometry interface. Rapid Commun Mass Spectrom 14:1377–1383
137. Benetton S, Kameoka J, Tan AM, Wachs T, Craighead H, Henion JD (2003) Chip-based P450 drug metabolism coupled to electrospray ionization-mass spectrometry detection. Anal Chem 75:6430–6436

138. Ma B, Zhang GH, Qin JH, Lin BC (2009) Characterization of drug metabolites and cytotoxicity assay simultaneously using an integrated microfluidic device. Lab Chip 9:232–238

139. Xue QF, Foret F, Dunayevskiy YM, Zavracky PM, McGruer NE, Karger BL (1997) Multichannel microchip electrospray mass spectrometry. Anal Chem 69:426–430

140. Ramsey RS, Ramsey JM (1997) Generating electrospray from microchip devices using electroosmotic pumping. Anal Chem 69(13):2617–2617, 69(6):1174–1174

141. Huikko K, Ostman P, Grigoras K, Tuomikoski S, Tiainen VM, Soininen A, Puolanne K, Manz A, Franssila S, Kostiainen R, Kotiaho T (2003) Poly(dimethylsiloxane) electrospray devices fabricated with diamond-like carbon-poly(dimethylsiloxane) coated SU-8 masters. Lab Chip 3:67–72

142. Rohner TC, Rossier JS, Girault HH (2001) Polymer microspray with an integrated thick-film microelectrode. Anal Chem 73:5353–5357

143. Sainiemi L, Nissila T, Jokinen V, Sikanen T, Kotiaho T, Kostiainen R, Ketola RA, Franssila S (2008) Fabrication and fluidic characterization of silicon micropillar array electrospray ionization chip. Sens Actuat B Chem 132:380–387

144. Dethy JM, Ackermann BL, Delatour C, Henion JD, Schultz GA (2003) Demonstration of direct bioanalysis of drugs in plasma using nanoelectrospray infusion from a silicon chip coupled with tandem mass spectrometry. Anal Chem 75:805–811

145. Arscott S, Le Gac S, Rolando C (2005) A polysilicon nanoelectrospray-mass spectrometry source based on a microfluidic capillary slot. Sens Actuat B Chem 106:741–749

146. Hoffmann P, Hausig U, Schulze P, Belder D (2007) Microfluidic glass chips with an integrated nanospray emitter for coupling to a mass spectrometer. Angew Chem Int Ed 46:4913–4916

147. Zheng YF, Li HF, Guo ZH, Lin JM, Cai ZW (2007) Chip-based CE coupled to a quadrupole TOF mass spectrometer for the analysis of a glycopeptide. Electrophoresis 28:1305–1311

148. Mao XL, Chu IK, Lin BC (2006) A sheath-flow nanoelectrospray interface of microchip electrophoresis MS for glycoprotein and glycopeptide analysis. Electrophoresis 27:5059–5067

149. Kameoka J, Orth R, Ilic B, Czaplewski D, Wachs T, Craighead HG (2002) An electrospray ionization source for integration with microfluidics. Anal Chem 74:5897–5901

150. Li FA, Wang CH, Her GR (2007) A sheathless poly(methyl methacrylate) chip-CE/MS interface fabricated using a wire-assisted epoxy-fixing method. Electrophoresis 28:1265–1273

151. Zhang BL, Foret F, Karger BL (2000) A microdevice with integrated liquid junction for facile peptide and protein analysis by capillary electrophoresis/electrospray mass spectrometry. Anal Chem 72:1015–1022

152. Razunguzwa TT, Lenke J, Timperman AT (2005) An electrokinetic/hydrodynamic flow microfluidic CE-ESI-MS interface utilizing a hydrodynamic flow restrictor for delivery of samples under low EOF conditions. Lab Chip 5:851–855

153. Xie J, Miao YN, Shih J, Tai YC, Lee TD (2005) Microfluidic platform for liquid chromatography-tandem mass spectrometry analyses of complex peptide mixtures. Anal Chem 77:6947–6953

154. Vollmer M, Horth P, Rozing G, Coute Y, Grimm R, Hochstrasser D, Sanchez JC (2006) Multi-dimensional HPLC/MS of the nucleolar proteome using HPLC-chip/MS. J Sep Sci 29:499–509

155. Tan AM, Benetton S, Henion JD (2003) Chip-based solid-phase extraction pretreatment for direct electrospray mass spectrometry analysis using an array of monolithic columns in a polymeric substrate. Anal Chem 75:5504–5511

156. Jo K, Heien ML, Thompson LB, Zhong M, Nuzzo RG, Sweedler JV (2007) Mass spectrometric imaging of peptide release from neuronal cells within microfluidic devices. Lab Chip 7:1454–1460

157. Willis RC (2006) Challenges for clinical diagnostic devices. Anal Chem 78:5261–5265

Chapter 2
Analysis of Herbicides on a Single C30 Bead via the Platform Combined Microfluidic Device with ESI-Q-TOF-MS

2.1 Introduction

Compare to the conventional chemistry and biology technologies, microfluidic device possesses the obvious advantages due to its small scale and high specific surface area. The microchannel, which has the similar dimension with cells, provides a powerful tool for the cell analysis. Mass spectrometry (MS), which is considered as a high sensitive detection method, is the best option for quantitative and qualitative analysis of chemical substance. The combination of these two techniques will play an important role in the study of exploring the physiological functions in organs or searching the essential unknown substance in cell reactions.

However, because the cells have very small sizes and flexible shapes, the content of the secretion and intracellular substances are extremely low. Therefore it's difficult to capture the cells in microfluidic channels and detect the secretion by MS directly. According to the current laboratory situation, we first built up an analysis platform combined the microfluidic devices with MS. A rigid solid-phase particle with the micro-size, which is easy to be captured and fixed, was introduced into the microchannel to simulate the cell. Trace content of compounds adsorbed on the solid-phase particle were detected by MS. Meanwhile, the developed method also provided an alternative option for the manipulation and capture of single particles, also the enrichment and detection of the environmental samples.

In this work, the herbicides were selected to be absorbed by the solid-phase particle. Herbicides are highly toxic when the content reaches a certain level. The increasing applications of herbicides during the last decades have resulted in the contamination of both soil and water, which result in a potential highly hazardous pollutant. Many studies on determination of herbicides in food and environmental samples have been reported with the conventional methods [1, 2]. Several groups have developed and validated methods for trace level herbicides determination in soil or food based on liquid chromatography [3] or gas chromatography [4], after liquid-liquid extraction [5], microwave-assisted extraction [6], solid-phase microextraction(SPME) [3], and in-tube SPME [7]. However, these traditional

H. Wei, *Studying Cell Metabolism and Cell Interactions Using Microfluidic Devices Coupled with Mass Spectrometry*, Springer Theses, DOI 10.1007/978-3-642-32359-1_2,
© Springer-Verlag Berlin Heidelberg 2013

methods are mostly time-consuming, and inevitably require complicated sample cleanup procedures. Additionally, when the samples are not easy to collect so that only very small amount of samples could be used for detecting, the conventional methods are lack of power.

In this chapter, we developed a simple, rapid, and economical approach to analysis the herbicides in the environmental samples. An analysis platform combined the microfluidic device with ESI-Q-TOF MS was firstly developed. Under the precise control of adsorption time and particle size, the single C30 particle was placed in the real sample extracted solutions to adsorb herbicides. The microchip was special designed to trap the single C30 particles. The high sensitive ESI-Q-TOF MS was employed to identify and qualify the eluate from the C30 particle. From the mass spectra obtained during the elution process, the accurate identification could be recognized for each target compound.

Very small amount of sample was required to complete the purification and detection procedures with the participation of the single C30 solid-phase particle. The parallel treatment processes could be carried out simultaneously on several C30 particles which were trapped in the microchannels. The operation time was greatly saved, and more information would be obtained in shorter time. The high-resolution mass spectrometry helped on the qualitative and semi-quantitative analysis of the target compounds. The reproducibility and recovery were confirmed, and a low detect limit could be reached. The microfluidic devices developed in this work, which were combined with MS detection on-line, is able to be applied in a variety of environmental samples.

2.2 Experimental Section

2.2.1 Reagents and Materials

Silicon wafers were purchased from Xilika Crystal polishing Material Co., Ltd. (Tianjin, China). Negtive photoresist (SU-8 2050), and developer were obtained from Microchem Corp. (Newton, MA, USA). Poly(dimethylsiloxane) (PDMS) and the curing agent were purchased from Dow Corning (Sylgard 184, Midland, MI, USA). The C30 silicone polymer-coated silica gel was obtained from GL scientific Corporation (Tokyo, Japan). The glass slides were purchased from Fisher Scientific (Pittsburgh, PA, USA). The terbutryn, simazine, propazine, ametryn, prometryn, and prometon selected as herbicides standards were obtained from Riedel-de Haën (Seelze, Germany). Methanol and acetonitrile (HPLC grade) were obtained from Fisher (New Jersey, USA).

A plasma Cleaner (PDC-32G, Harrick Plasma, Ithaca, NY, USA) was employed for oxygen plasma treatment. A syringe pump (KDS100, kdScientific, Holliston, MA, USA) was used to inject the eluting solution in accurate rate. A fluorescence microscope (Leica DMI 4000 B, Wetzlar, Germany) equipped with a charge coupled

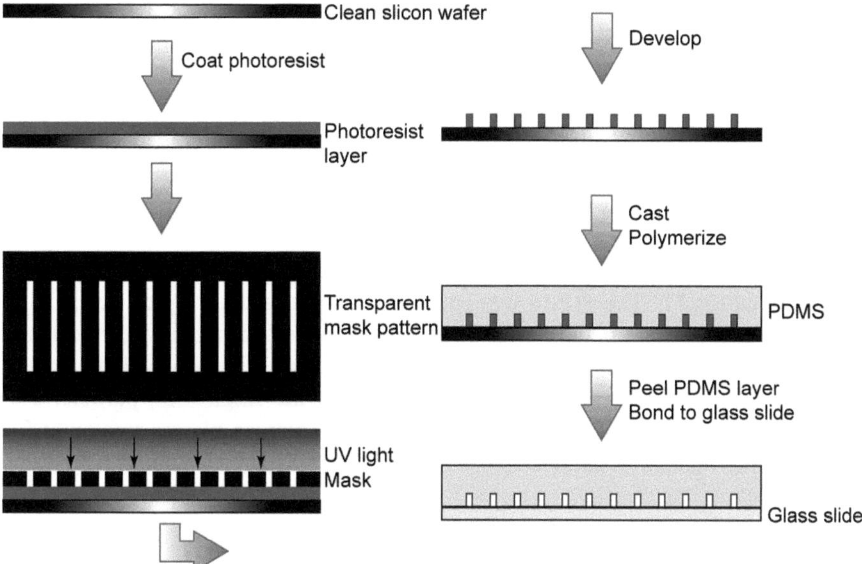

Fig. 2.1 Illustration of the fabrication procedures of microfluidic device using soft lithography

device (CCD) camera was used to observe and photograph the microfluidic devices. The diameters of each microstructure were manually measured by Leica Application Suite, LAS V2.7. The mass spectrometry detection was performed with a Bruker micrOTOF-Q mass spectrometer (Bruker Daltonics Inc., Billerica, MA, USA), and the obtained mass spectra were performed in the positive mode. MS image analysis was performed by DataAnalysis, the Bruker homemade software. A 500µL syringe was obtained from Hamilton, Bonaduz AG, Switzerland.

2.2.2 Fabrication of Microfluidic Devices

The standard soft lithography technique [8] was applied to fabricate the microfluidic devices used for single particles analysis by the poly(dimethylsiloxane) (PDMS). As shown in Fig. 2.1, the fabrication procedure of the microfluidic device is illustrated. The piranha solution (30% H_2O_2: 98% sulphuric acid = 1:3, v/v) was used to modify and clean the surface of the silicon wafer. After washed and dried, the wafer was spin-coated by the negative photoresists SU-8 2050 at 3,000 rpm (revolutions per minute) to form a 50 µm thick film. After the pre-bake, the wafer was exposed under the UV light, in order to transfer the pattern of the transparent mask onto the silicon wafer. The mask with the pattern designed and draw by Adobe Illustrator was printed by laser photo-typesetting. As shown in Fig. 2.1, the black area was light-proof, and the white area was transparent. The UV-light went through the white area and started the polyreaction of the negative photoresist, so that the reacted area cannot be washed off by the developer. After being developed by the SU-8 developer, the mold that carried a relief of the desired microstructure was generated.

Prior to use in soft lithography, the patterned silicon wafer was exposed to 1H,1H,2H,2H-perfluorooctyl trichlorosilane vapor in a vacuum desiccator for 2 h in order to prevent adhesion between the cured PDMS and the mold. A 10:1 pre-mixed PDMS prepolymer was prepared according to the manufacturer instructions, degassed in a vacuum chamber for 1 h and then poured into the mold and cured in a 70°C oven for 2 h. The PDMS was cut from the mold with a surgical scalpel and then carefully peeled off the mold. The channel inlets and outlets were punched by a shape flat-tip syringe needle. An extra hole was also punched out in the middle of the microchannel as required to contain the single C30 particle. The channels were sealed with a glass slide after being oxygen plasma treatment for 90 s. The microchannels obtained were 15 mm long, 800 μm wide and 50 μm deep.

2.2.3 Sample Preparation

The herbicides standards stock solutions were prepared in methanol at the concentration of 1.0 mg/mL. The working solutions were prepared by diluting the stock solutions in methanol. The mixed standard solution was obtained by mixing six herbicides standard solution by the same volume, and stored after being mixed homogenously. All the solutions were stored at 4°C, and filtrated by the water-phase membrane with 0.45 μm holes before experiments, in order to prevent the block in the microchannels and the contamination in the ESI source. The six herbicides monitor ion and structures are shown in Table 2.1.

The carrots and potatoes were chosen as the tuber vegetable samples for analysis. First, the vegetable samples were cut into small cubes and homogenized by a food blender. Then 100 mL acetonitrile was added into 300 g of the homogenized sample, and deposited in an ultrasonic washer 10 min for fully contacting and extracting. The ventilating nitrogen gas was applied on the extracted solution to accelerate the solvent evaporation until it was dry. Five hundred microliters acetonitrile was added to dissolve the solute to get the constant volume. The condensed extract was collected in a 1.5 mL centrifuge tube after filtered by a 0.45 μm microporous member, and then centrifuged 5 min at 2,000 rpm. Afterwards the C30 beads were placed in the tube containing the supernatant for 5 min to absorb the herbicides.

2.2.4 Single Particle Adsorption

Silicone polymer-coated silica gel modified with C30 alkyl chains was normally used as packing materials for reversed-phase liquid chromatography. The surface was coated by porous silica gel with a homogeneous silicone polymer film, and

Table 2.1 Herbicides monitor ion and structure [14]

Analyte	Monitor ion (m/z)	Structure
Prometryn	242[M+H]$^+$	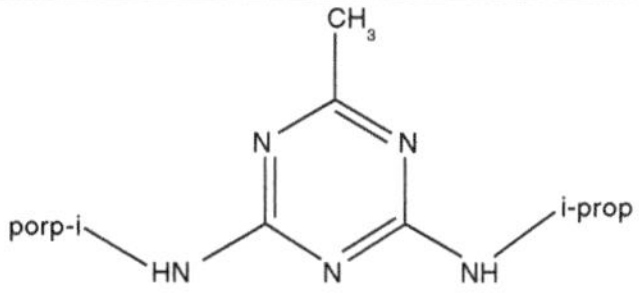
Terbutryn	242[M+H]$^+$	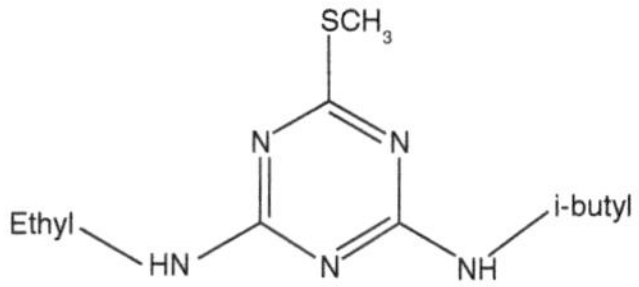
Propazine	230[M+H]$^+$	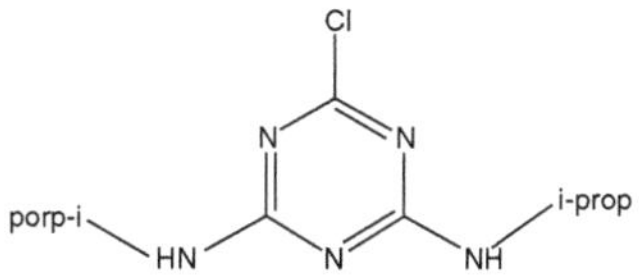
Ametryn	228[M+H]$^+$	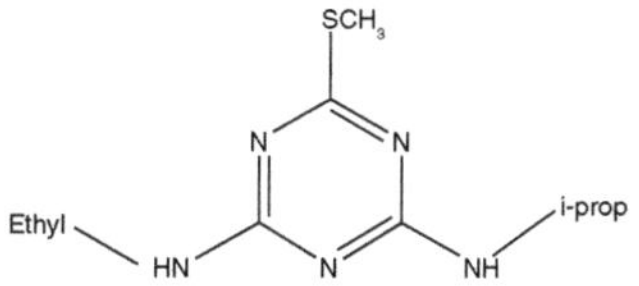
Prometon	226[M+H]$^+$	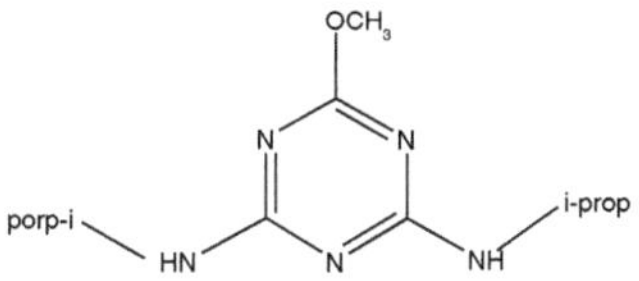
Simazine	202[M+H]$^+$	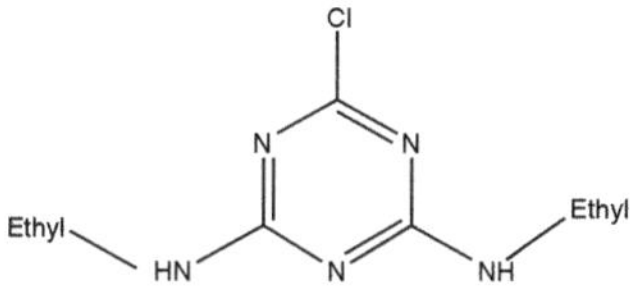

Reprinted from Ref. [14], with permission from Elsevier

thereafter modifying the coating polymer with C30 alkyl groups, which shows strong resistance of alkali-like organic porous polymeric materials. C30 bonded reversed-phase silica was selected as the experimental object not only for its high capability for adsorption but also for its size which is suitable for manually manipulation. The average exterior diameter of the C30 beads was roughly 400 μm (Fig. 2.2), which was much larger than the height and width of the channel which was fabricated to trap the C30 beads.

The C30 beads were washed by deionized water and methanol for three times before the adsorption. The washing procedure was carried out by vortexing, centrifuging, and removing the supernatant in order. The target C30 beads were selected by tweezers with a sharp tip, and transferred into the determined solutions. The diameters of the C30 beads chosen in the experiments are measured under the microscope with the error less than 30 μm.

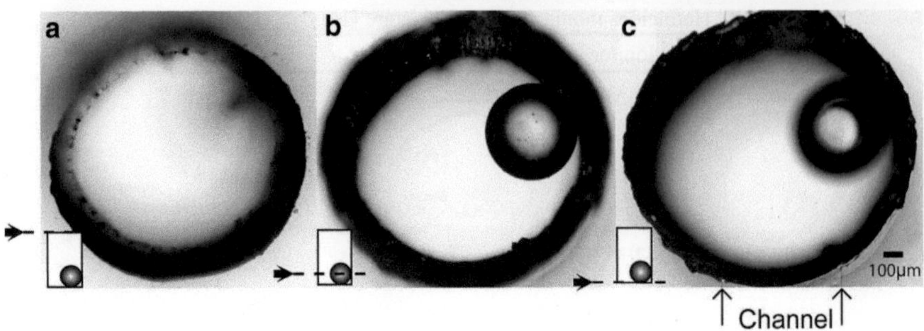

Fig. 2.2 Microscopic images of a C30 bead the microfluidic channel. (**a**) *Top of the channel* was in focus, and the C30 bead cannot be observed. (**b**) The *center* of the bead was in focus so that the bead can be measured accurately. (**c**) Focus was continual turn down to the *bottom* of the hole and the boundary of channel could be observed clearly (Reprinted from Ref. [14], with permission from Elsevier)

2.2.5 Single Particle Manipulation and Elution

After the sample pretreatment, individual C30 bead was transferred from the determined solution into the hole which was punched in the middle of the microchannel, and pushed to the bottom of the chip by a flat-tip syringe needle very carefully. A stainless steel solid column with the diameter of 1.5 mm and length of 10 mm was used to seal the hole. The hole was completely sealed because the PDMS is elastic, and the diameter of the stainless steel column is a little larger than the hole on the microchannel. Proper space was left for the C30 bead and the steel column should not be pushed to the very down bottom, to prevent the C30 bead from being crushed. The stainless steel columns were washed by the deionized water and methanol in ultrasonic washer.

The washing solvents for eluting herbicides were the mixture of methanol–water at the flow rate of 30 µL/h. The methanol and water was mixed with the volume ratio of 2:3, and complemented by 1 ‰ acetic acid to enhance the ionization. After elution, the column was removed, and the C30 bead was taken out. Both of the microchannel and the C30 bonded silica were reusable.

2.2.6 Mass Spectrometry Settings

The electron spray ionization-quadrupole-time of flight mass spectrometry (ESI-Q-TOF) was used for all experiments. The inlet capillary was heated at 200°C. A 30 µL/h flow rate was applied for sample infusion. The coaxial nebulizer N_2 gas flow around the ESI emitter was used to assist generation of ion. The ESI-Q-TOF MS was externally calibrated by tunemix (Agilent, USA) at the mode of positive method in the range of 50–1,500 m/z.

The determination for each single C30 bead only cost 5 min to identify and semi-quantify the chemicals. Only 2.5 µL solvents were required for each C30 bead analysis. The limit of quantification of the mass spectra was calculated from ten times experiments with ESI-Q-TOF MS.

2.3 Results and Discussion

2.3.1 Combination of Microfluidic Device and Mass Spectrum

In recent years, many efforts have been made to combine microfluidic devieces with a mass spectrum, especially via the reconstruction of the electrospray interface. The most commonly adopted approaches are mainly through the fused-silica capillaries or nanospray needles [9–11], and the external sprayers also have been described [12, 13]. However, these methods are complicatedly designed and manipulated, with a high consumption in solvents and time.

In this work, we connected the microchip with ESI ion source through a simple capillary suite, which is shown in Fig. 2.3. The capillary with the polytetrafluoroethylene (PTFE) cannulas on both ends matched both of the joint for connecting ESI ion source and the injector. The capillary was also inserted in the inlet and outlet of the microchannels on PDMS microchip to connect the eluting flow.

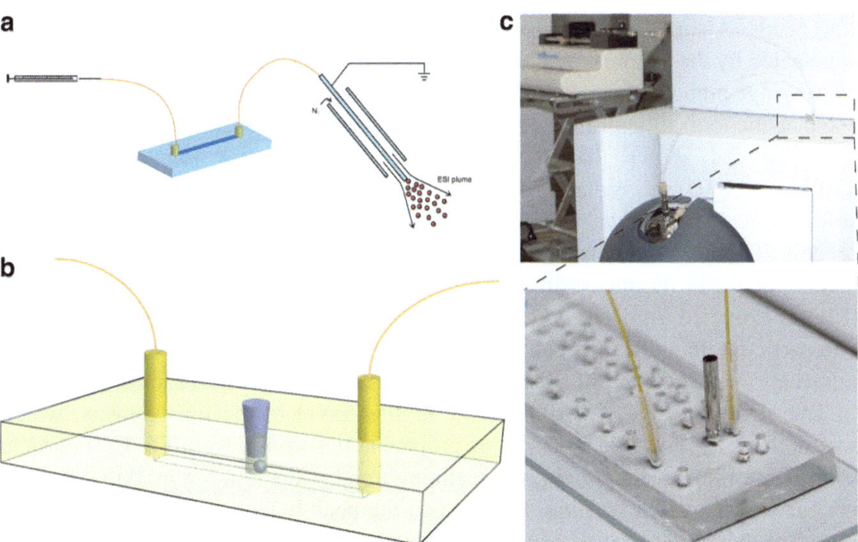

Fig. 2.3 Microfluidic device combined with ESI-Q-TOF-MS. (**a**) Scheme of coupling the microchip and Q-TOF mass spectrum together by capillaries. (**b**) Schematic diagram shows the trapped single C30 bead inside microchannel using the hole on the *top side*. (**c**) Image of the whole device of coupling microchip and MS. The microfluidic chip is magnified (Reprinted from Ref. [14], with permission from Elsevier)

2.3.2 Optimization of the Sample Preparation

Before applying the C30 single particle to the vegetable samples, it is necessary to evaluate the capacity of C30 bead to absorb the herbicides. The single C30 bead was exposed to the herbicides solutions after washing and activation. Then the C30 bead attached by herbicides was placed in the microfluidic chip which was connected to the MS, eluted by the mobile solvents, and detected by the ESI-Q-TOF MS.

The time for the C30 beads absorbing the herbicides was optimized. Twenty-five C30 beads were added into a 100 ppm propazine solution at the same time. Every group including five beads was taken out at 5, 20, 60, and 120 min, while the last five beads were remained in the propazine solution for 12 h. Individual bead was placed in the microchip device, and analyte was analyzed as previously described. The mass spectra obtained from the five groups beads were compared, which result in that the longer absorption process didn't significantly increase the amount of the absorbed analyte. In order to shorten the time consumption, a 5 min absorption time was adopted.

2.3.3 Propazine Analysis by Single C30 Bead

In order to evaluate the efficiency of the microfluidic devices by single bead analysis, the propazine standard solution with series concentrations were analyzed by single C30 beads. A single C30 bead was placed in the hole in the microchannel which was sealed by the stainless steel column afterwards. The mobile phase was injected by a syringe pump at the rate of 30 μL/h. 2‰ acetic acid was added in the solvent to enhance the ionization of the analytes under positive method in mass spectrometry.

Figure 2.4 shows that the intensity of the monitor ion peak of m/z 230.1128($\pm$0.05) was reducing during the eluting process. Due to the DataAnalysis software made by Bruker, various curves could be extracted during the monitoring process. The total ion chromatogram (TIC) indicated the total intensity of all the defined peaks in the measuring range. The curve of extracted ion chromatogram (EIC) is extracted from TIC with the desired m/z value, according to which the trend of a particular monitor ion peak is traced (Fig. 2.4a). Image of EIC helps to get the information for the trend of aimed ion, in condition of interruptive ion existing.

The arrow on Fig. 2.4a showed the highest point of the ion intensity, which was defined as time 0. The curve of propazine desorption from the C30 particle solid-phase indicated an obviously declining trend at the first 5 min. After 25 min, the curve displayed a flat line, from which the peak height of m/z 230 was lower than 500, which was much weaker compared to the time 0. The consumption of the eluting solvent was 2.5 μL for a single time elution to identify a chemical and 12.5 μL solvent required for attaining an equilibrium condition. As a result, considering the time spent on sample preparation, the identification could be achieved in less than 10 min, with less than 2.5 μL organic solvent consumed.

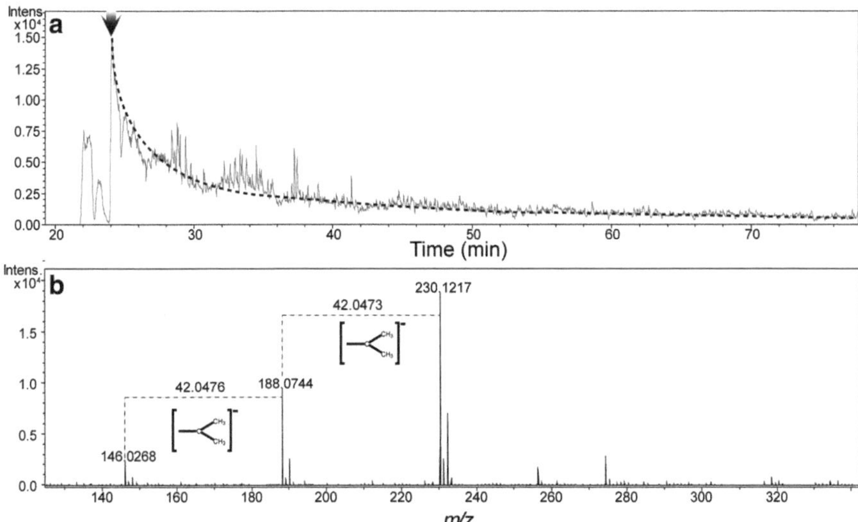

Fig. 2.4 Propazine desorption process from a single C30 bead. (a) Elution curve of propazine desorption process from the solid phase. The *dash line* shows the trend of EIC image. (b) Mass spectra of propazine desorbed from the C30 silica, which was obtained at the right time the *arrowhead* indicated (Reprinted from Ref. [14], with permission from Elsevier)

After performed with data analysis, according to the MS image obtained at the time point that the highest signal was observed (Fig. 2.4b), the monitor ion peak of propazine were clearly present in the spectrum. The peak of m/z 188 indicated the fragment that propazine losing an isopropyl, while the peak of m/z 146 proved the fragment losing two isopropyls.

To evaluate the possibility of the semi-quantified analysis by the single C30 bead with microfluidic devices, the herbicides standard solutions with series concentrations were performed. Figure 2.5 shows the relationship between the intensity of the peak m/z 230 and the concentrations of the propazine solutions. The stock solution of propazine standard was diluted 10, 100, 1,000 times separately. The propazine standard solutions were analyzed by the single C30 bead as the procedures described before. The calibration curve obtained on plotting the peak areas was linear in the range of 1–1,000 ppm and the fitting formula was $Y = 1212.97 + 1368.09X$ while the $R^2 = 0.9938$. The limits of detection (LOD) and limits of quantification (LOQ) were obtained by detecting the blank sample via single C30 beads 20 times and calculated until S/N reached three and ten respectively. The results showed that the LOD was 0.11 ppm, while the LOQ was 0.36 ppm. The reproducibility was in the range of 3–11%, which was determined by detecting the serial standard solutions and blank control.

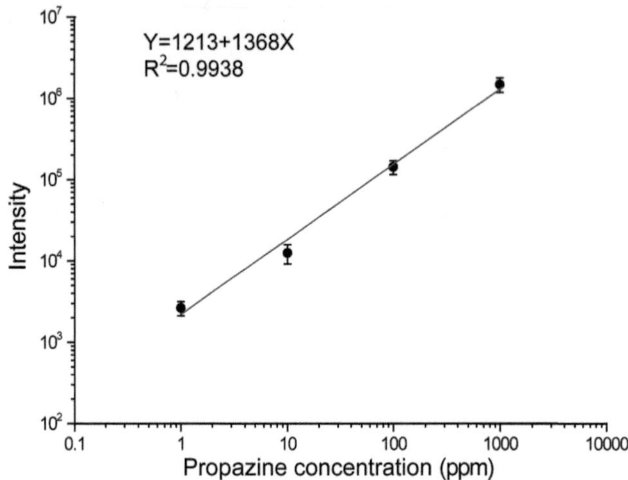

Fig. 2.5 The relationship between the intensity of the peak m/z 230 and the concentration of the propazine solution (Reprinted from Ref. [14], with permission from Elsevier)

2.3.4   Herbicides Mixture Elution and Identification

In order to evaluate the performance of the single C30 particle analysis with microfluidic devices on chemicals mixture, six herbicides standard solutions were mixed homogenously with the same volume. The similar structures of these six herbicides decided their similar abilities of getting protons, which directly affect the detection sensitivity by mass spectrometry. Figure 2.6 shows the mass spectra of the C30 bead attached herbicides mixture. The curve of Fig. 2.6a indicates the eluting process of the herbicides desorbing from the solid phase. Images of EIC including m/z 242.1545($\pm$0.05), m/z 230.1219($\pm$0.05), m/z 228.1376($\pm$0.05), m/z 226.1781($\pm$0.05), m/z 202.1016($\pm$0.05), which were extracted from TIC, shows the intensity trend of five ion peaks. In the mass spectra image of the herbicides desorbed from the C30 bead, which was gained at the point with the highest intensity, five monitor ion peaks indicated the existence of five chemicals at least. Moreover, the further structure identification was achieved by MS/MS method with the extra power of 70 eV.

Tandem mass spectrometry (MS/MS) connected with the database for identification is an essential tool for the identification of chemicals which are difficult to distinguish, and an assistant method in order to confirm the molecular structures. The focused monitor ion was selected and transferred into quadrupole mass analyzer. In the MS mode, the quadrupole mass analyzer allows all the ions pass through. When analyzing the isomeric compounds, extra energy is loaded on the quadrupole mass part. As a result the target ions are selected and transferred into the time-of-flight analyzer while the other ions with different m/z are blocked outside. Extra power generates new fragments, which assist to define the structure of chemicals.

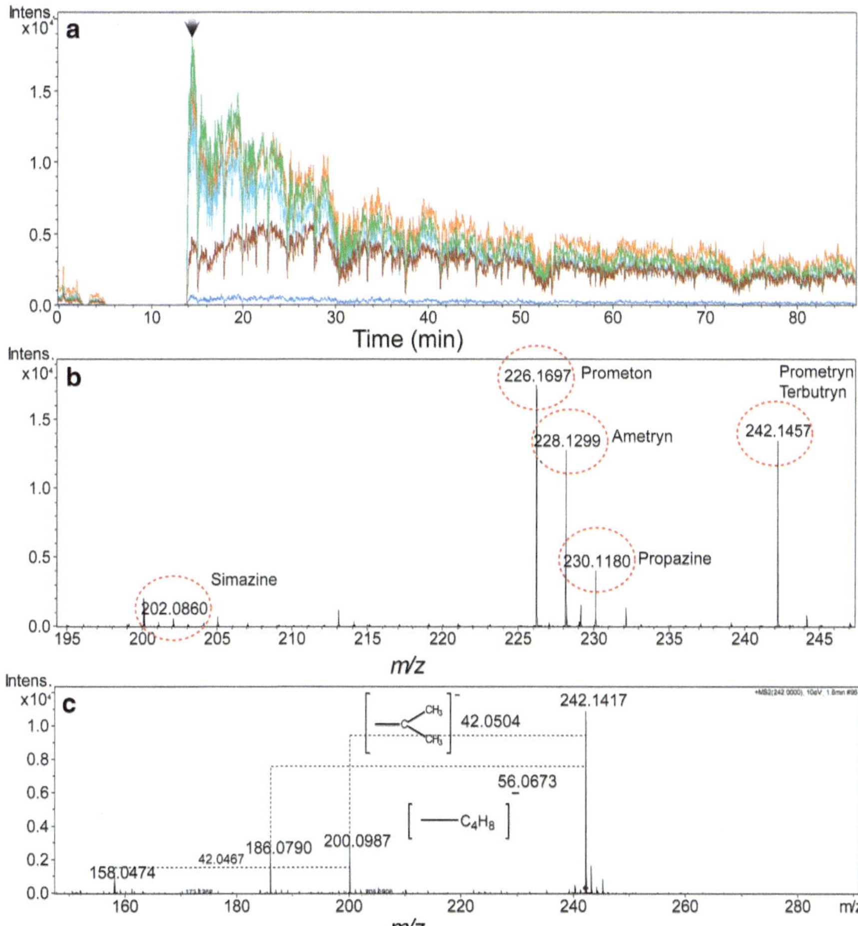

Fig. 2.6 Herbicides mixture desorption process from a single C30 bead. (**a**) Elution curve of the desorption process. Different *colored curves* reveal five monitor ion peaks separately. (**b**) Mass spectra of the herbicides desorbed from the C30 silica, which was obtained at the time point that the *arrowhead* indicated. (**c**) Monitor ion of *m/z* 242 (Reprinted from Ref. [14], with permission from Elsevier)

In order to identify the prometryn and terbutryn, monitor ion of *m/z* 242 (Fig. 2.6c) was analyzed under the MS/MS mode. Prometryn and terbutryn, which are isomeric compounds, revealed the same monitor ion peak as *m/z* 242 in the mass spectra. The peak of *m/z* 158 and 200 indicated the existence of prometryn, as the result of losing one and two isopropyl; meanwhile the peak of *m/z* 186 was the fragment generated from terbutryn, losing an isobutyl.

Table 2.2 The detection results of herbicides in vegetable samples [14]

Intensity / [M+H]⁺ Samples	202	226	228	230	242
Extract of carrot	412	908	869	1833	---
Extract of potato	---	255	592	534	---

Reprinted from Ref. [14], with permission from Elsevier

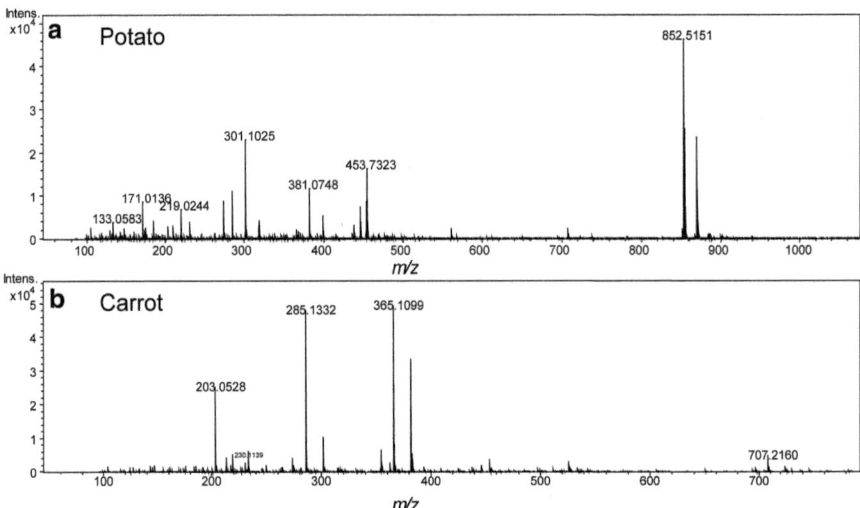

Fig. 2.7 Mass spectra of vegetable samples. Acetonitrile was used to extract the herbicides in vegetable homogenized samples (Reprinted from Ref. [14], with permission from Elsevier)

2.3.5 *Analysis of Vegetable Samples*

The tuber vegetables were collected from local market. The vegetable samples were homogenized and extracted by organic solvents. After pretreatment, the extracted solutions were measured following the established approach by the single C30 beads. The results are listed in Table 2.2, in which the value is the average of ten times replicates.

The mass spectra are shown in Fig. 2.7. In both carrot and potato sample, four monitor peaks of herbicides can be marked, whose intensity were much higher than the baseline. The further identification was accomplished under the MS/MS mode. 10 eV extra energy was applied on the extract ion m/z 202, m/z 226, m/z 228, m/z 230 separately to get the fragments. The mass spectra of fragments were shown in Fig. 2.8.

The propazine in the vegetable samples was determined and semi-quantified. According to the standard curve, the content of propazine in carrot is 0.45 ppm,

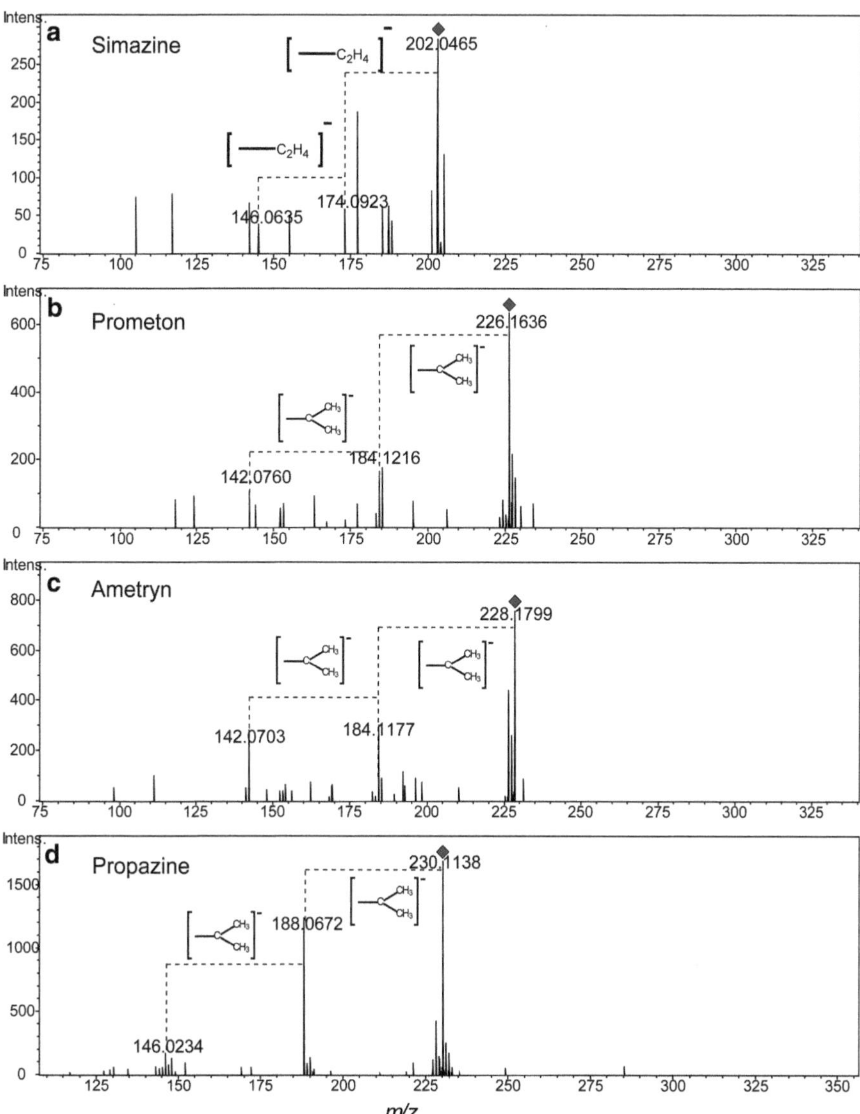

Fig. 2.8 MS/MS spectra of herbicides detected in carrot sample (Reprinted from Ref. [14], with permission from Elsevier)

which could be calculated as 0.75 μg/kg in the carrot sample. And the content of propazine in potato is out of the linear range, which totally met the regulations for food. As a result, none of the propazine in tuber vegetables we purchased reached the legislation of EPA, EU, or Japanese MRLs, which is 0.05 mg/kg at least.

2.4   Conclusion

In this work, we developed an analysis platform combined the microfluidic device and ESI-Q-TOF-MS together to manipulate a single C30 bead for the herbicides detection. Compared to conventional pretreatment methods, such as SPME, the application of single C30 bead has greatly simplified the procedure, by avoiding the usage of separation solid-phase column and thus the extra steps of removing the interfering components. Microfluidic device was introduced to manipulate the single C30 bead for its small dimension. The trace level chemicals eluted from the single C30 bead were detected and identified by mass spectrum, which showed accurate and clear characteristic peaks for aimed compounds. The detection and semi-quantity of all herbicides could be completed in 5 min without separation, which greatly shortened time and saved solvents. The investigations into single particle analysis efficiency showed that the microfluidic device could reduce the analysis time, and solvent consumption, and could be applied on analysis of a varied of environmental samples.

References

1. Papadopoulos N, Gikas E, Zalidis G, Tsarbopoulos A (2009) Simultaneous determination of herbicide terbuthylazine and its major hydroxy and dealkylated metabolites in Typha latifolia L. Wetland plant using SPE and HPLC-DAD. J Liq Chromatogr Relat Technol 32:2975–2992
2. Sambe H, Hoshina K, Haginaka J (2007) Molecularly imprinted polymers for triazine herbicides prepared by multi-step swelling and polymerization method – their application to the determination of methylthiotriazine herbicides in river water. J Chromatogr A 1152:130–137
3. Turiel E, Tadeo JL, Martin-Esteban A (2007) Molecularly imprinted polymeric fibers for solid-phase microextraction. Anal Chem 79:3099–3104
4. Avramides EJ, Gkatsos S (2007) A multiresidue method for the determination of insecticides and triazine herbicides in fresh and processed olives. J Agric Food Chem 55:561–565
5. Aramendia MA, Borau V, Lafont F, Marinas A, Marinas JM, Moreno JM, Urbano FJ (2007) Determination of herbicide residues in olive oil by gas chromatography-tandem mass spectrometry. Food Chem 105:855–861
6. Cheng JH, Liu M, Zhang XY, Ding L, Yu Y, Wang XQ, Jin HY, Zhang HQ (2007) Determination of triazine herbicides in sheep liver by microwave-assisted extraction and high performance liquid chromatography. Anal Chim Acta 590:34–39
7. Lord HL (2007) Strategies for interfacing solid-phase microextraction with liquid chromatography. J Chromatogr A 1152:2–13
8. Unger MA, Chou HP, Thorsen T, Scherer A, Quake SR (2000) Monolithic microfabricated valves and pumps by multilayer soft lithography. Science 288:113–116
9. Wheeler AR, Freire SLS (2006) Proteome-on-a-chip: mirage, or on the horizon? Lab Chip 6:1415–1423
10. Zamfir AD, Bindila L, Lion N, Allen M, Girault HH, Peter-Katalinic J (2005) Chip electrospray mass spectrometry for carbohydrate analysis. Electrophoresis 26:3650–3673
11. Foret F, Kusy P (2006) Microfluidics for multiplexed MS analysis. Electrophoresis 27:4877–4887

12. Brivio M, Tas NR, Goedbloed MH, Gardeniers HJGE, Verboom W, van den Berg A, Reinhoudt DN (2005) A MALDI-chip integrated system with a monitoring window. Lab Chip 5:378–381
13. Ostman P, Marttila SJ, Kotiaho T, Franssila S, Kostiainen R (2004) Microchip atmospheric pressure chemical ionization source for mass spectrometry. Anal Chem 76:6659–6664
14. Wei H, Li H, Lin J-M (2009) Analysis of herbicides on a single C30 bead via a microfluidic device combined with electrospray ionization quadrupole time-of-flight mass spectrometer. J Chromatogr A 1216:9134–9142

Chapter 3
Monitoring of Glutamate Release from Neuronal Cell Based on the Analysis Platform Combining the Microfluidic Devices with ESI-Q-TOF MS

3.1 Introduction

In order to further explore the law of life activities, the attentions should be focused on the cells, which are the basic unit of life. The essence of life activities is the cell behavior. In recent years, cell analysis is playing an important role in the research of intercellular substrate, cell secretion, and metabolites. It's of major significance especially in the early diagnosis, treatment, and medicine screening of the major diseases, as well as the cell physiological and pathological processes.

In Chap. 2, the analysis platform combining the microfluidic devices with mass spectrometry was successfully established and evaluated as a qualitative and quantitative detection method for trace substances. Based on this analysis platform, the microfluidic devices were improved to trap and fix the cells in microchannels, in order to further study the cells secretions. To demonstrate the feasibility of developed cell analysis platform, the medicine screening for preventing the Alzheimer's disease (AD) was selected as a model, which is a degenerative disorder of the central nervous system and the most common form of dementia affecting the elderly. The characteristic pathological features found of AD include intra-neuronal neurofibrillary tangles and the deposition of extracellular amyloid plaques containing amyloid-β peptide (Aβ). PC12 (a cell line derived from a pheochromocytoma of the rat adrenal medulla) cells are normally employed in experiments to simulate the habits of neuronal cells. Under the *in vivo* condition, the extracellular concentrations of glutamate are maintained at a low level due to the mechanism of glutamatergic neurotransmission [1]. One of the main biochemical functions of Aβ42 peptide-induced neurotoxicity is interrupting and damaging the glutamate cyclic transmission. As a result of the interruption, glutamate is released into the extracellular matrix.

Hereby the analysis system combining microfluidic devices with ESI-Q-TOF MS was described to manipulate the extracellular environment of cells, collect the resulting secretion products released by the cells, and characterize the products with a sensitive analyser for medical screening tests or rapid diagnosis. Poly-*L*-lysine

H. Wei, *Studying Cell Metabolism and Cell Interactions Using Microfluidic Devices Coupled with Mass Spectrometry*, Springer Theses, DOI 10.1007/978-3-642-32359-1_3, © Springer-Verlag Berlin Heidelberg 2013

was coated in the microchannels to culture PC12 cells. A multi-channel miniature extraction chip (MEC) was integrated to remove salts and protein interferences. ESI-Q-TOF MS was employed to perform the semi-quantitative and highly sensitive qualitative analysis. The protective effect of carnosine against Aβ42-induced neurotoxicity was evaluated under different conditions in microchannels in parallel. The PC12 cells secretion analysis was accomplished in 5 min with the solvent consumption of only 2.5 μL. The combination analysis platform of microfluidic devices and mass spectrometry has significant potential for the analysis of cellular secretions, as well as for medical screening tests and for the diagnosis of specific diseases.

3.2   Experimental Section

3.2.1   Reagents and Materials

Negative photoresist (SU-8 2050), and the developer were obtained from Microchem Corp. (Newton, MA, USA). Silicon wafers were purchased from Xilika Crystal polishing Material Co., Ltd. (Tianjin, China). Poly-dimethylsiloxane (PDMS) and the curing agent were purchased from Dow Corning (Sylgard 184, Midland, MI, USA). Poly (ethylene glycol) diacrylate (PEG-DA, MW 700) and 2-hydroxy-2-methylpropiophenone (HMPP) photoinitiator were obtained from Aldrich Chemical Co. (Milwaukee, WI, USA). The PEG precursor solution containing 3% photoinitiator was diluted by water with the same volume and then stored at 4°C before use. 3-(Trichlorosilyl)propyl methacrylate (TPM) was purchased from Fluka Chemicals (Milwaukee, WI, USA). Poly-*L*-lysine coated glass slides and carnosine were purchased from Sigma Chemical Co. (St. Louis, MO, USA). Methanol and acetonitrile (HPLC grade) were obtained from Fisher (New Jersey, USA). Aβ42 was obtained from BioSource International (Camarillo, CA, USA). A live/dead viability/cytotoxicity assay kit (Invitrogen, CA, USA) was obtained for viability tests on encapsulated cells. The packaging materials for pretreatment were obtained from SPE (solid phase extraction) Bond Elut Plexa cartridges (polymer beads) (Varian, Melbourne, VIC, Australia).

A syringe pump (KDS100, kdScientific, Holliston, MA, USA) was used to deliver eluting solutions at accurate rates. A plasma cleaner (PDC-32G, Harrick Plasma, Ithaca, NY, USA) was employed for oxygen plasma treatments. A fluorescence microscope (Leica DMI 4000 B, Wetzlar, Germany) equipped with a CCD camera (Leica DFC 300 FX, Wetzlar, Germany) was carried out to observe and obtain images of the microfluidic devices. The dimensions of each microstructure were manually measured with Leica Application Suite, LAS V2.7. The mass spectrometry detection was performed with a Bruker microTOF-Q mass spectrometer (Bruker Daltonics Inc., Billerica, MA, USA), and the mass spectra were obtained under positive mode. MS images analysis was performed with DataAnalysis, a proprietary software program provided by Bruker. A 500 μL syringe was purchased from Hamilton (Bonaduz AG, Switzerland).

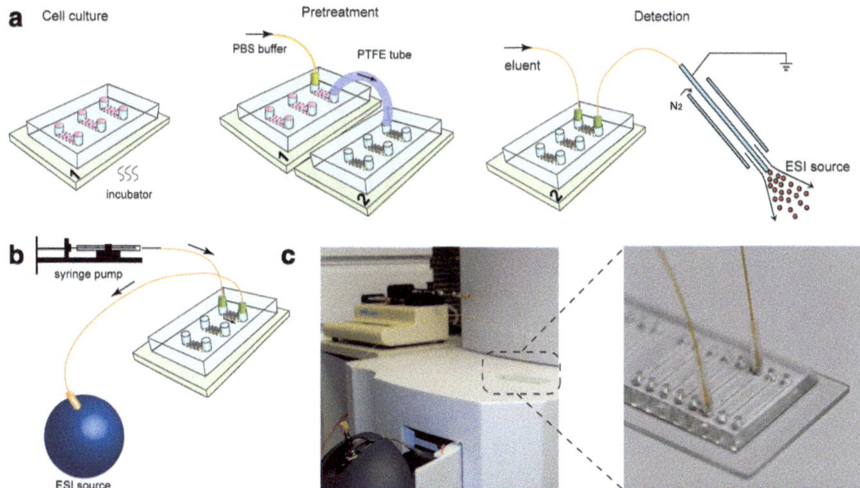

Fig. 3.1 Illustration of integrated microfluidic devices combined with ESI-Q-TOF MS. (**a**) Schematic diagram of the analysis procedures: PC12 cells culture, secretion pretreatment, and detection by MS. (**b**) Scheme of the integration devices. (**c**) Image of the microfluidic device and MS. An *enlarged view* of the microchip is shown on the *right* (Reproduced from Ref. [4], with permission from The Royal Society of Chemistry)

3.2.2 *Fabrication of Microfluidic Devices*

The desired microstructures was produced by poly-dimethylsiloxane (PDMS) following the standard soft lithography techniques. Briefly, the mold used for generating microchips of cell culture and MEC were fabricated by spincoating the negative photoresist SU-8 2050, at 3,000 rpm (50 μm thick film) and 1,800 rpm (80 μm thick film) separately on silicon wafers which were prior cleaned by piranha solution. Then the master molds were exposed and developed following the lithography technique to transfer the designed patterns on the silicon wafers. The 10:1 premixed PDMS prepolymer was prepared, degassed, and poured on the mold, before placing in a 70°C oven for 2 h. The cured PDMS was removed from the mold with a surgical scalpel and then carefully peeled off. A flat-tip syringe needle was used to punched the channel inlets and outlets. The fabricated PDMS microchips were bonded on the cleaned glass slides by the oxygen plasma treatment for 90 s. The microchannels obtained for PC12 cell cultures were 1 mm wide, 15 mm long, and 50 μm deep, while the dimensions of the channels for the pretreatment MEC were 1 mm wide, 15 mm long, and 80 μm deep.

Figure 3.1 shows the scheme of the integrated analysis platform. As described in the last chapter, the microfluidic device was connected with ESI ion source *via* a PTFE tube. The microchannels sealed with the poly-*L*-lysine coated glass slides were employed for the PC12 cell cultures, while the channels sealed with normal glass slides were carried out for fabricating the pretreatment MEC. Firstly, the cells

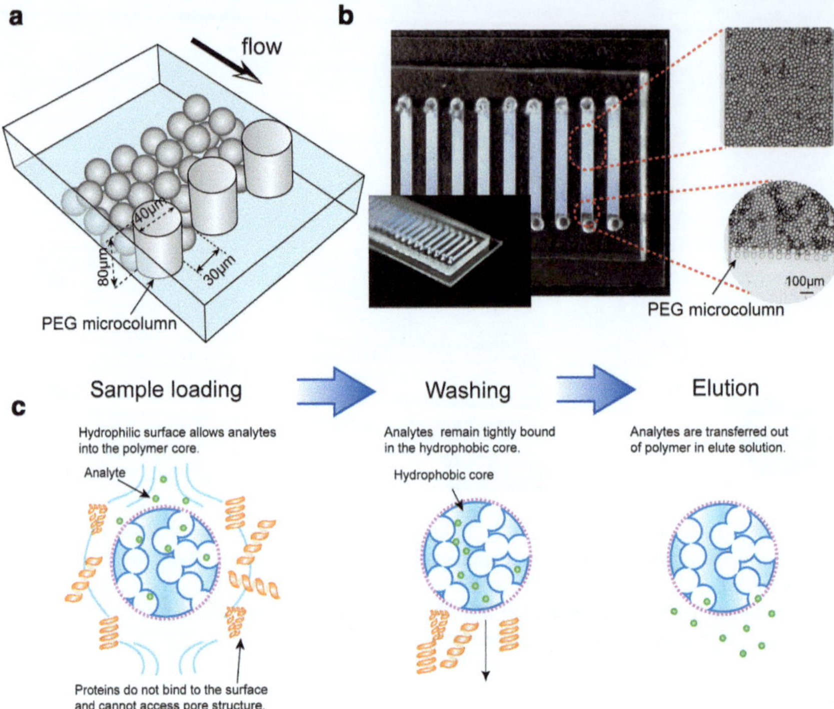

Fig. 3.2 Scheme of the MEC for pretreatment. (**a**) Illustration of the MEC fabrication in a microchannel. (**b**) The *image* shows the MEC and the PEG microcolumns under microschop. (**c**) The pretreatment procedure including sample loading, washing, and elution (Reproduced from Ref. [4], with permission from The Royal Society of Chemistry)

were cultured in channel 1 until they propagated to certain concentration. Then required chemicals were applied to the cells for a while, the cells secretion was collected and pretreated through channel 2. In the end, channel 2 was connected to the mass spectrometry to achieve the qualitive and quantitive analysis of the eluted compounds.

3.2.3   Fabrication of Integrated MEC for Sample Pretreatment

A MEC was integrated in the microfluidic devices for the pretreatment of glutamate collected from cellular supernatants. The layout of the integrated MEC in the microfluidic device is shown in Fig. 3.2.

A dam composed by PEG microcolumns array was generated inside the chip by UV-initiated polymerization. The packaging material was obtained from Bond Elut Plexa cartridges. It provided a highly hydrophilic surface which excluded

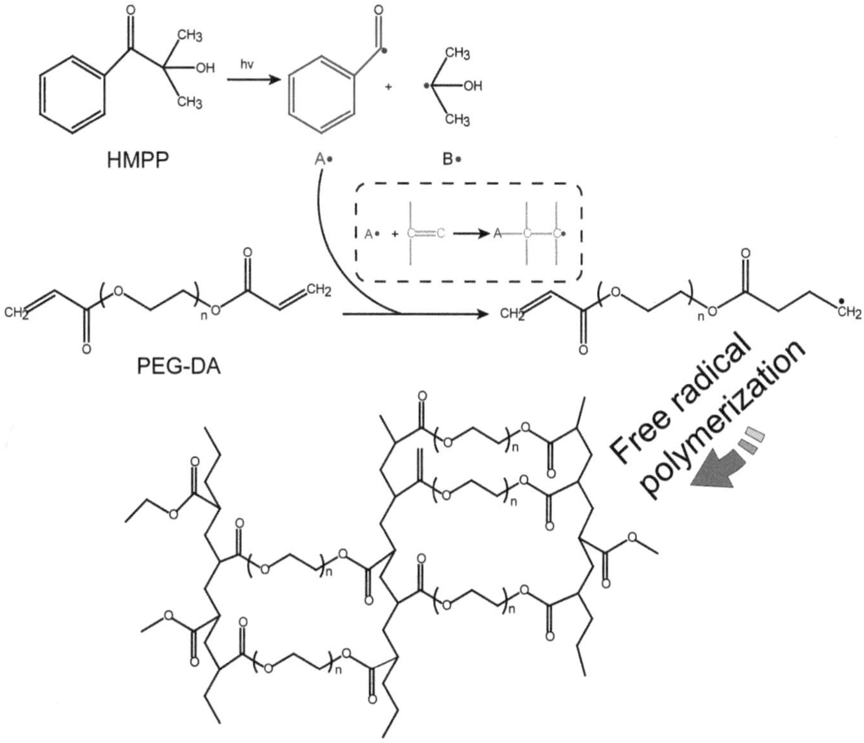

Fig. 3.3 Scheme of the polymerization function of building the PDG-DA three-dimensional network structure with HMPP as the photoinitiator

proteins and lipophilic interferences from binding. As shown in Fig. 3.2c, the solid phase which was polymerized based on a polarity gradient inside the bead, allowed the target analyte binding at the core. The space between the neighbouring PEG microcolumns was 30 μm, which is smaller than the external diameter of the packaging material as 45 μm.

The 2-hydroxy-2-methyl-1-phenyl propanone (HMPP), which photofragments to yield a highly reactive methyl radical, was selected as the photoinitiator to induce the PEG hydrogel polymerization. As shown in Fig. 3.3, Free radicals initiate the polymerization of a copolymer network by attacking the carbon-carbon double bonds (C=C) of the acrylate groups of PEG-DA. A three-dimensional insoluble polymer network was formed in the polymerization which resulted in branched and crosslinked structures. The polymerization procedure is as follows: The PEG-DA was injected into the MEC microchannels. After the polymerization under UV light, the unpolymerized compounds were flushed away, and the PEG hydrogel microstructures array was obtained (Fig. 3.4).

The inner glass surfaces of the microchannels were modified with a monolayer of 3-(trichlorosilyl)propyl methacrylate (TPM), in order to enhance the adhesion

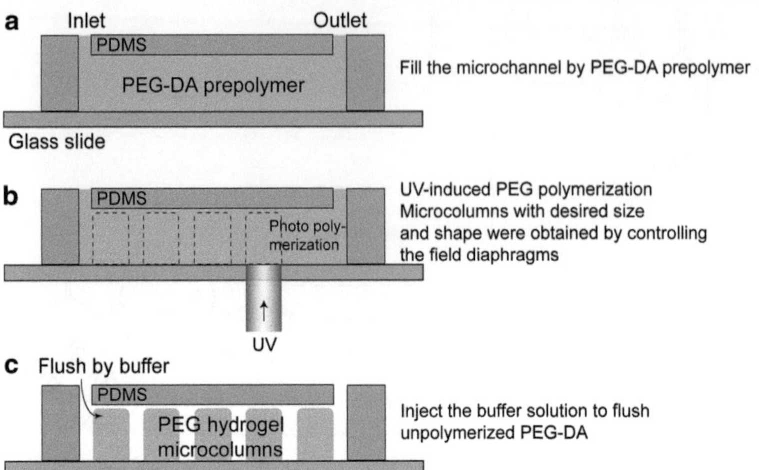

Fig. 3.4 Illustration of the procedures to fabricate the PEG hydrogel microstructures in MEC microchannel

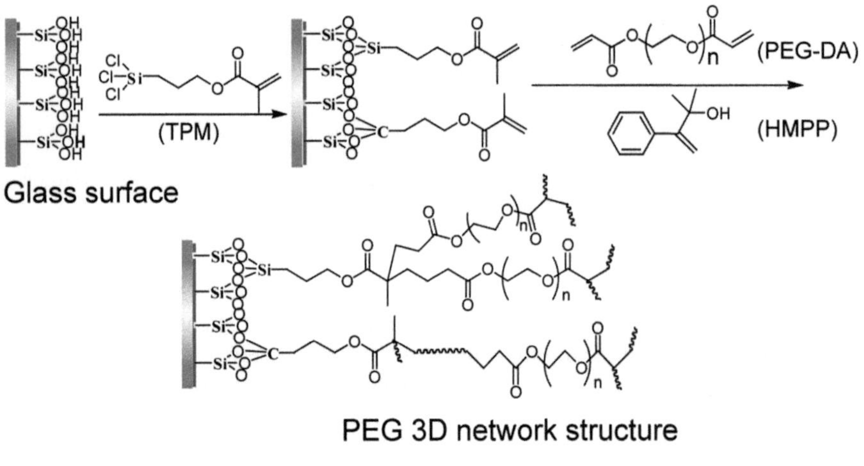

Fig. 3.5 TPM was modified to enhance the adhesion between the PEG hydrogel microstructures and the glass slides

between the PEG hydrogel microstructures and the glass slides (Fig. 3.5). A 1% (v/v) TPM solution in paraffin oil was injected into the microchannels and maintained for 10 min to achieve the modification. The microchip was rinsed by ethanol, placed into an oven at 95°C for 30 min, and dried under nitrogen gas flow afterwards. Subsequent photopolymerization covalently linked the gel to the glass slide *via* the silane-coupling agent.

The procedures of fabrication the PEG microcolumns inside the microchannels are as follows: The PEG hydrogel precursor solution consisting of 5% (w/w)

PEG-DA and 0.05% (w/w) 2-hydroxy-2-methylpropiophenone photoinitiator in deionized water was injected into the microchannel. The fluorescence microscope with an external fluorescence light source was employed to photopolymerize the PEG precursor inside the microchannels. The UV light beam (340–380 nm) from the external fluorescence light source was focused on the PEG precursor inside the microchannels. A 50 ms exposure with UV light for photopolymerization was applied by a computer-controlled shutter on the microscope. Field diaphragms and objectives with a magnification of 40× were used to control the illuminated area inside the microchannels. The distance of the adjacent edges of the neighbouring PEG microstructures could be controlled manually up to 2 μm. The fabrication of the dam in each MEC channel took less than 1 min. Unpolymerized hydrogel precursor solution was removed by washing the channels with deionized water at 100 μL/min via a syringe pump for 10 min.

3.2.4 Mass Spectrometry Settings

The electron spray ionization-quadrupole-time of flight mass spectrometry (ESI-Q-TOF) was used for all experiments. The inlet capillary was heated at 200°C. A 30 μL/h flow rate was applied for sample infusion. The coaxial nebulizer N_2 gas flow around the ESI emitter was used to assist generation of ion. The ESI-Q-TOF MS was externally calibrated by tunemix (Agilent, USA) at the mode of positive method in the range of 50–1,500 m/z. Multiple reaction monitoring (MRM) in the positive ion mode was performed under a collision energy of 10 eV to detect the fragmentation m/z 130.1. All the provided values are average of ten measurements with ESI-Q-TOF MS.

3.2.5 Biological Experiments

3.2.5.1 Preparation of β-Amyloid Peptide

The stock solution of Aβ42 was dissolved in sterile, double-distilled water at the concentration of 1 mg/mL, and stored in aliquots of 50 μL at −20°C. Prior to use, the Aβ42 peptide was incubated at 37°C for 1 week to allow aggregation. Afterwards the stock solution was diluted to the desired experimental concentrations.

3.2.5.2 PC12 Cells Culture

The differentiated PC12 cells were donated by the Department of Biology, Tsinghua University (Beijing). Cells were grown in pH 7.4 growth medium consisting of Dulbecco's Modified Eagle's Medium (DMEM) supplemented with 10% horse serum, 5% heat-inactivated fetal calf serum, 100 U/mL penicillin G, and 100 U/mL streptomycin, and cultured at 37°C and 5% CO_2 in a humidified incubator. Subculture was taken place by trypsinizing by the 0.25% trypsin solution.

3.2.5.3   PC12 Cell Immobilization in Microfluidic Channels

The glass surface in microchannels was coated with poly-*L*-lysine as described before to help the PC12 cells to adhere. Prior to seeding the cells, the microfluidic devices were sterilized with 75% ethanol for 30 s and then rinsed thoroughly with deionized water. The integrated microfluidic device was sterilized by ultraviolet radiation on a laminar flow cabinet in a clean room for 30 min and stored inside until the performance of the cell seeding.

Before seeding the cells, the cell culture channels were initially filled with phosphate buffered saline (PBS) buffer to remove the air. The PC12 cells suspension was prepared at the concentration of 2×10^5 cells/mL. The cells suspension should be gently pipette before injected into each microchannel to prevent cells sedimentation. The extra suspension in the outlets was carefully removed by a pipette with thin top. Cells adhered to the bottom of the channel after culturing for 2 h. Culture medium was dropped on the inlet, due to the hydrophobic surface of PDMS microchip, a droplet was formed to cover the inlet, and the culture medium automatically flow towards the outlet because of the gravity. The culture medium was refreshed every 8 h. Experiments were carried out 24 h after cells were seeded.

3.2.5.4   Drug Exposure

In our experiments, drugs were dissolved in sterile purified water prior to their dilution with PBS. Drugs solutions prepared by PBS buffer were applied to PC12 cells to avoid complications from the complex culture medium solution. The control experiment was simultaneously carried out by culturing cells only in PBS buffer. In the experiments which demonstrated the protective effect of carnosine against Aβ42-induced neurotoxicity, after exposed in the carnosine solution for 18 h, the PC12 cells were exposed to Aβ42 for 24 h.

3.2.5.5   Cell Viability Test

It was reported by many works that the cells can maintain their viability for several days inside microchannels by controlling the microenvironment. Calcein acetoxy-methyl ester (Calcein AM, 4 μM in PBS buffer) and ethidium homodimer-1 in the presence of DNA (EthD-1, 8 μM in PBS buffer) from the live/dead viability/cytotoxicity assay kit for mammalian cells was applied to determine the viability of PC12 cells. The polyanionic dye calcein is well retained within living cells, producing an intense uniform green fluorescence (ex/em ∼495 nm/∼515 nm).

EthD-1 enters cells with damaged membranes and undergoes a 40-fold enhancement of fluorescence upon binding to nucleic acids, thereby producing a bright red fluorescence in dead cells (ex/em ∼495 nm/∼635 nm) [2]. The change of the green fluorescent intensity indicates the change of the cell viability, while the fluorescence colour changing from green to red indicates cell cytotoxicity. The viability stains were introduced into the inlets of the microchannels and maintained for 20 min to incubate at 37°C. According to the manufacturer's instructions, the prepared aqueous working solutions were valid within 1 day.

3.3 Results and Discussion

In this work, we focused on collecting and determining the secretions released from PC12 cells based on the established analysis platform, in order to investigate the protection effect of chemicals against Aβ42-induced neurotoxicity. Moreover, experiments with different stimulation protocols were carried out to characterize the various released secretion conditions, which could help the future studies on the therapeutic effects of drugs designed to treat Alzheimer's disease.

Neuronal cells play an important role in the correct mechanism of glutamatergic neurotransmission, especially for maintaining extracellular concentrations of glutamate at a low level. PC12 cells are cells derived from a pheochromocytoma of the rat adrenal medulla, and are often used as a model of neuronal cells. A process including the release, uptake, and metabolism of glutamate and glutamine is taking place in a cycle of transmission in PC12 cells (Fig. 3.6) [1]. Damaging the glutamate cyclic transmission is a main biochemical property of Aβ42 peptide-induced neurotoxicity. As a result, glutamate is released into extracellular space, in which process the underlying mechanisms are still unclearly. Carnosine (β-alanyl-histidine) is a major constituent of the brain and a putative neurotransmitter. Because of the abundant functions such as anti-oxidant, anti-glycation, and anti-aging, the

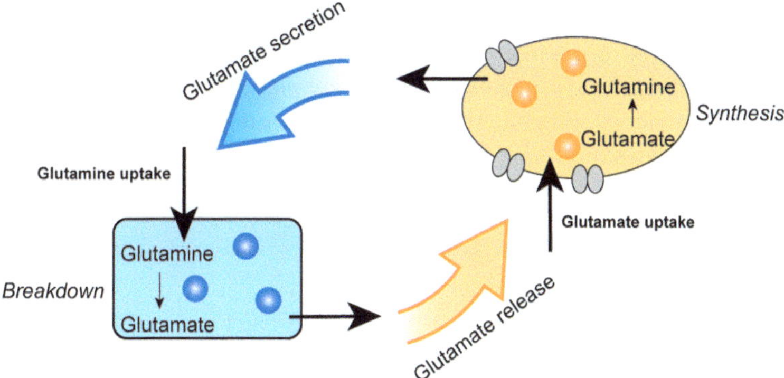

Fig. 3.6 Glutamate's release, uptake, and metabolism in a PC12 cell

carnosine is supposed to have a protective effect on AD. Therefore, there is a great interest to find out the protection effect of carnosine against Aβ42-induced neurotoxicity, and further contribute in the medical screening against AD.

The developed cell analysis platform allowed to culture the cells and analysis the secretions by mass spectrometry, after extracting the secretions of interest from complex biological matrices. The microfluidic devices for cell culture and secretion pretreatment were fabricated with parallel multiple microchannels integrated in two chips respectively. The cell culture chip and the pretreatment MEC were designed separately, because the complex culture medium was applied for cell culture in the microchip, which could harm the MEC device and thus disrupt the MS measurements. Furthermore, the MEC device could be prepared during the incubation of PC12 cells on the cell culture chip, which also shorten the manipulation time consumption.

3.3.1 The MEC for Pretreatment

The MEC for cells secretion pretreatment was prepared by polymerizing PEG hydrogel microstructures inside the microchannels. The order of magnitude of the packaging material beads was in the range of 30–50 μm, which determined the distances between the microstructures inside the microchannel should be small enough to trap the beads. However, fabrication of such small and precise microstructures was limited by current soft lithography techniques. As reported, the hydrogel microstructures could be easily obtained by a fluorescence microscope with illumination control. Moreover, hydrogel microstructures with different sizes could be fabricated by shifting to different magnifications, and the distance of the adjacent edges of the hydrogel microstructures could be adjusted manually. Based on this technique, a microdam was fabricated in the channel to trap packaging material beads, with the distances between neighboring microcolumns much smaller than the diameters of the beads (Fig. 3.2).

In this work, the fabricated PEG microcolumn was obtained with the diameter of 40 μm, while the distance between two PEG microstructures was less than 30 μm, which was achieved manually. The PEG microcolumns were arranged into two lines. The microstructures in the second line were placed facing the interface of the adjacent ones in the first line, in order to efficiently trap without blocking the flow.

3.3.2 Evaluation of the MEC

The binding capacity of the MEC was evaluated, in order to determine the methodological feasibility of quantification analysis. The standard glutamate solutions with known concentrations was directly injected into the pretreatment channel.

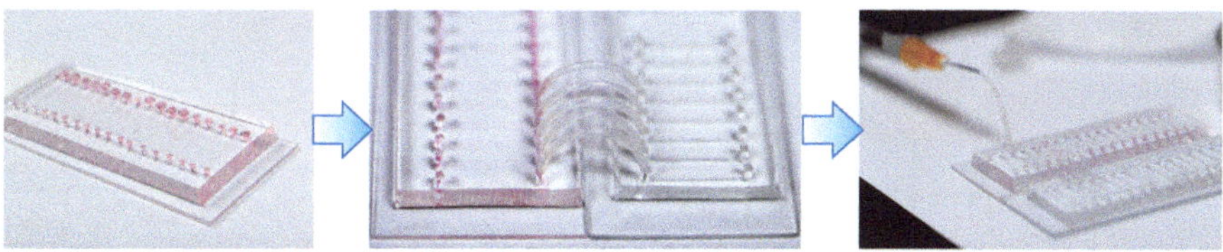

Fig. 3.7 Scheme of the process of removing the sample solutions from cell culture microchannels into the MEC microchannels

After being isolated from the Bond Elut Plexa cartridges, the packaging materials were flushed with methanol and suspended by vortexing. The suspension was injected into the prepared microchannels containing a PEG dam to pack the MEC. Porior to loading the samples, the MEC was conditioned by washing with 100 μL methanol followed by 100 μL water. After applying samples, the MEC was washed by 100 μL 5% methanol, and then eluted by 5% ammoniated methanol (v/v) solution. The elution was directly injected into the mass spectrometer for detection. All the pretreatment procedures with the MEC were performed by a syringe pump with the flow rate of 10 μL/min, except in the case of sample loading procedure, which was performed with a flow rate of 2 μL/min.

The glutamate solution was diluted with PBS buffer to the concentrations as 0.01, 0.05, 0.1, 0.5, and 1 μg/mL. First, the standard glutamate solutions were injected into the cell culture microchannel to get the same volume as the desired volume of real sample. Then the solution was removed into the pretreatment MEC through a PTFE tube which connected the corresponding microchannels in the cell culture microchip and MEC. PBS buffer was injected from the inlet of the cell culture microchannel to push the glutamate solutions into the MEC microchannel (Fig. 3.7). MS detection was carried out after elution steps.

Figure 3.8 shows the relationship between the intensity of the m/z 147.1 peak and the concentration of the glutamate standard solutions. The result of the analysis confirmed that reducing the glutamate concentration from 1 to 0.01 μg/mL caused the signal intensity increased 100 times. The calibration curve obtained on plotting the peak areas was linear in the range of 0.01–1 μg/mL, with a fitting formula of Y $= 9087X + 636$ and an R^2 of 0.9961. The precision of the approach was evaluated by repeating each experiment five times. The relative standard deviation (RSD) was varied from 7 to 20% at higher concentrations, while the lowest concentration showed a RSD of 24%. The RSD results revealed the acceptable reproducibility of the experiments at a higher concentration range. The limit of quantitation (LOQ) was calculated based on measurement of the blank solvent ten times, which turns out as 2 μg/L. Cell culture and stimulation in the integrated microfluidic device. The evaluation results that the integrated MEC allows to extract a solution with a concentration of 1 μg/mL glutamate from the cell culture microchannel, which is much higher than the concentration that can be released by PC12 cells.

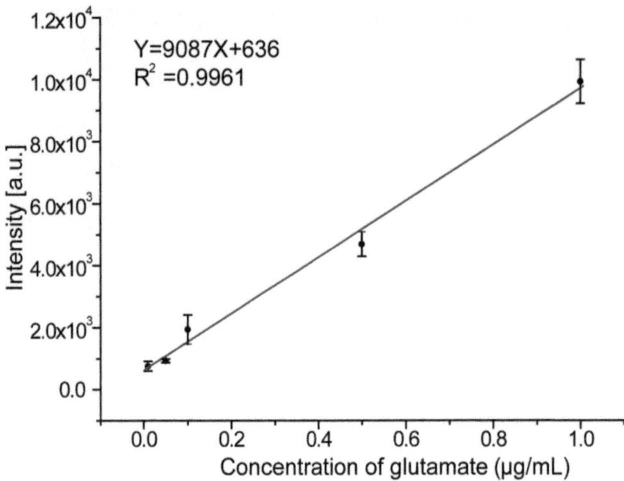

Fig. 3.8 The relationship between the intensity of the *m/z* 147.1 peak and the concentration of the glutamate standard solutions (Reproduced from Ref. [4], with permission from The Royal Society of Chemistry)

3.3.3   Cell Culture and Drug Exposure on Microfluidic Chips

After seeding in the prepared cell culture microchannels, the PC12 cells were cultured in regular growth medium for 24 h in order to promote the adhesion to the glass slide and improve cell vitality. In previous reported work, continuous perfusion of medium was normally applied to prevent the cells culture device from drying out [3]. However, the overnight injection by a syringe pump was required in the continuous perfusion approach, which also demanded a separate heating system to maintain the culture temperature, and the consumption of a large amount of culture medium. In this work, in order to provide adequate nutrient content and avoid evaporation in the culture incubator, 8 µL culture medium was pipetted on the inlet to form a droplet over the port. Because of the liquid pressure equilibrium between the inlet and outlet, both of the ports were covered by the culture medium. Due to the hydrophobicity of PDMS, the droplets covering the neighboring ports will not mix with each other, which prevented the contaminations between the channels. The culture medium was refreshed every 8 h to counteract the evaporation. The same protocols were performed when carnosine and Aβ42 solutions were applied. Therefore, only a small amount of medium was consumped to culture the cells, compared to the consumption in the 96-well method. This is especially valuable in the case of stimulation chemicals, which can be very expensive.

The sterilized PTFE tubes were employed to connect the cell culture channels with corresponding MEC channels. After being exposed to the drug solutions, the supernatant in the cell culture microchannel was transferred to the MEC microchannels by injecting PBS buffer through the inlets of the cell culture microchannels.

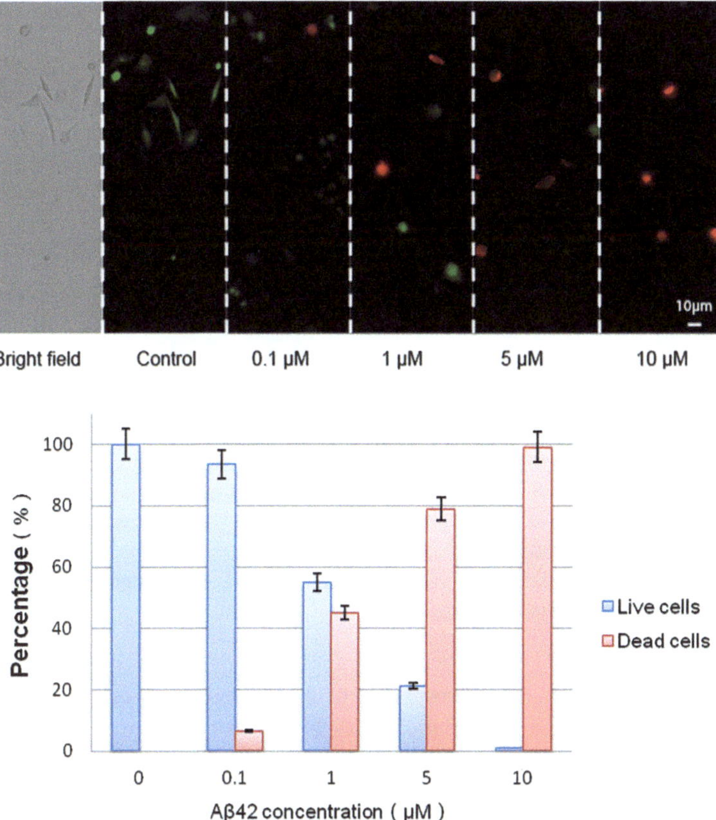

Fig. 3.9 Cells viability under Aβ42 solutions with different concentrations. *The upper images* are cells fluorescent photos after stained by the live/dead viability/cytotoxicity assay kits. *The lower figure* shows the statistics of the cells viability

3.3.4   Monitoring Aβ42-Induced Glutamate Release

Prepared Aβ42 solutions were applied to cultured cells in which PC12 cells had adhered to the glass surface and been cultured for 24 h. The cell suspension was diluted with growth medium to 1×10^5 cells/mL before seeding in to the microchannels, in order to get more sufficient contact of each cell when a shorter stimulation process was adopted.

The neurotoxicity of Aβ42 is shown in Fig. 3.9, the Aβ42 solution with concentration of 0.1, 1, 5, and 10 µM were applied to the PC12 cells cultured in microfluidic channels. The viability of cells was tested bylive/dead viability/cytotoxicity

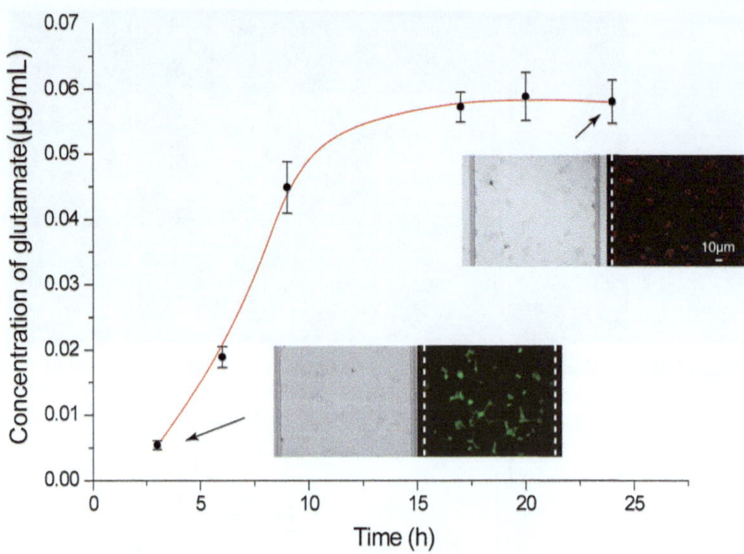

Fig. 3.10 Time-dependent glutamate release induced by Aβ42 neurotoxicity (Reproduced from Ref. [4], with permission from The Royal Society of Chemistry)

assay kits to investigate the neurotoxicity of Aβ42. The cells viability and survival rate declined when a Aβ42 solution with higher concentration was applied on the cells.

5 µM Aβ42 in PBS solution was applied to cultured cells to monitor the time-dependence of Aβ42-induced glutamate release. Glutamate release was monitored at 3, 6, 9, 17, 20 and 24 h. For each exposure time the experiments were carried out in five replicate microchannels. Figure 3.10 demonstrates the dependance of glutamate release on the duration of Aβ42 exposure.

The glutamate release turned to be constant after an exposure time of 13 h, which was much shorter than the time in conventional biological experiments. Additionally, the viability of cells was studied. The activity of cells was significantly higher after an exposure time of 3 h, while after a 24 h exposure, the cell shape was damaged and the cells totally died. These results confirmed the feasibility of the integrated microfluidic device for cell secretion analysis and low-cost, rapid medical screening tests.

One important characteristic of the microscale devices is the highly increased surface-to-volume ratio, as well as the faster diffusion process of chemicals among the immobilized cells. Moreover, the chemical and time consumption were greatly reduced in this approach. The comparison between the different methods is shown in Table 3.1.

Table 3.1 Comparison of conventional bioanalytical methods with our integrated microfluidic device method as applied to cell secretion analysis

Method	Time	Chemical consumption
Integrated microfluidic devices combined with ESI-TOF-MS detection		
Drug exposure in microchannels	22 h	1.5 μL
Pretreatment by MEC	10 min	100 μL
Secretions analysis by ESI-TOF-MS	5 min	2.5 μL
Regular biological method with HPLC-ECD detection [5]		
Drug exposure in 96-well plates	42 h	400 μL
Regular pretreatment	40 min	not given
Secretions analysis by HPLC-ECD	10 min	7,500 μL
Regular biological method with enzyme-luminescence detection [6]		
preparation of culture plates	1 day	200 μL
Drug exposure in 96-well plates (dopamine release)	1 h	10 μL
Secretions analysis by enzyme-luminescence assay	10 min	290 μL

Reproduced from Ref. [4], with permission from The Royal Society of Chemistry

3.3.5   *Effect of Carnosine Against the Aβ42-Induced Neurotoxicity*

The carnosine solutions used to protect the PC12 cells in the Aβ42-induced neurotoxicity was prepared as 0.1, 0.5, 1, 5 and 10 mM in PBS buffer. Prior to the exposure in the 5 μM Aβ42 solution for 24 h, the carnosine solutions were applied to PC12 cells for 18 h. The experiments were carried out on the PC12 cells after 24 h culture in microchannels for, due to the higher viability in the cells eugenic growth period.

Three control experiments were carried out simultaneously to show the carnosine protection effect on PC12 cells. First, to monitor the growing status of PC12 cells in PBS buffer, PC12 cells were cultured only in PBS buffer for 42 h as a blank control experiment. Second, in order to eliminate the contribution of carnosine in the release of glutamate, PC12 cells were cultured in the solution containing 5 mM carnosine for 42 h. Last, for the purpose of investigating the protection effect of carnosine against Aβ42-induced neurotoxicity, PC12 cells were exposed to 5 μM Aβ42 for 24 h after the cultivation process in PBS buffer for 18 h. The drug solutions were reflashed every 8 h as described above. All experiments were repeated five times to demonstrate the reproducibility.

The tendency of glutamate release after being exposed in different drug solutions is shown in Fig. 3.11. The PC12 cells were exposed in the Aβ42 solution with and without pretreatment with various carnosine solutions. It should be noted that during the process of the PC12 cells secretions collection, the possible existence of additional hydrophobic chemicals released by PC12 cells would also be adsorbed in MEC and eluted into the MS for detection later. However, only the peak of m/z 147.1 which was monitored as the glutamate peak during the detection process. It revealed that PC12 cells released most of their glutamate intracellular which were

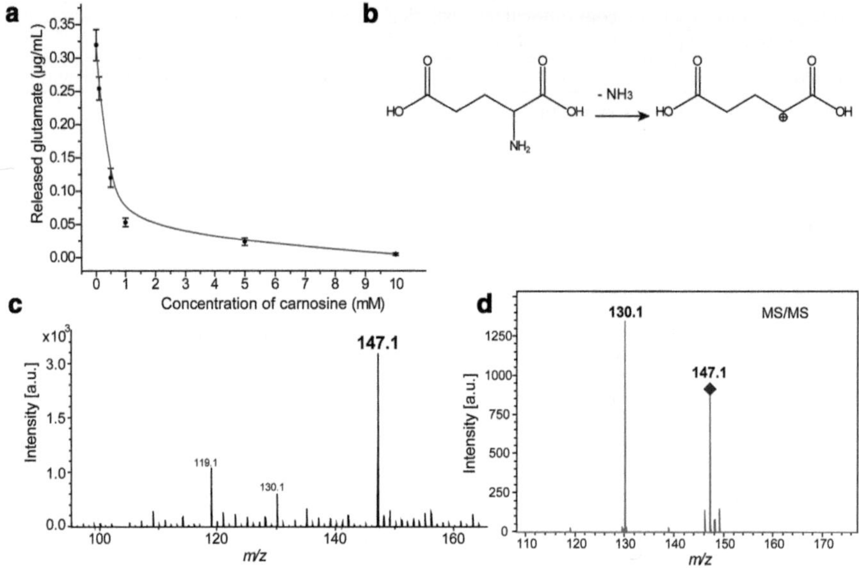

Fig. 3.11 (**a**) Different glutamate release from PC12 cells under various protection protocols after exposure to Aβ42. (**b**) Mass spectra of glutamate in the secretion after pretreatment by MEC. (**c**) Mechanism formula when the glutamate lost an amino group. (**d**) Mass spectra obtained under MRM mode (Reproduced from Ref. [4], with permission from The Royal Society of Chemistry)

exposed to 5 μM Aβ42 solution for 24 h. The PC12 cells which were applied by carnosine solutions with higher concentrations were observed releasing less glutamate after exposure to Aβ42. This is in agreement with theorized protection afforded by carnosine toward nerve cells against Aβ42 neurotoxicity.

There was only little glutamate release determined from the secretion collections which were obtained from PC12 cells cultured with PBS buffer for 42 h. While neither glutamate nor carnosine molecular ion peaks was found from the mass spectra obtained from PC12 cells cultured with carnosine solution, which demonstrated that the glutamate release was caused by Aβ42-induced neurotoxicity (mass spectra not shown). These results also revealed that the carnosine was taken up by PC12 cells, and that carnosine plays a role in the protection of the glutamate cycle in PC12 cells.

The mass spectra of the secretion obtained after pretreatment in MEC are shown in Fig. 3.11c. The target substance in PC12 cells secretion was identified by the mass spectra obtained within the mass range of m/z 100–200. A good response was shown of glutamate with $[M+H]^+=147.1$ by ESI-Q-TOF MS detection in the positive mode. The further confirmation was carried out by MRM on m/z 147.1 in positive ion mode, in which the amino group could be broken down under an extra power of 10 eV and a characteristic fragment at 130.1 was found.

The viability of cells was determined by the live/dead viability/cytotoxicity assay kits to investigate the protection effect of carnosine. Table 3.2 compares the different

Table 3.2 The survival of PC12 cells in different drug exposure protocols

Protocol	Cell survival (%)
PBS buffer (42 h)	95.7 ± 2.4
Carnosine (5 mM, 42 h)	93.5 ± 3.5
PBS buffer (18 h)[a]	7.1 ± 2.3
Carnosine (0.1 mM, 18 h)[a]	28.9 ± 4.3
Carnosine (0.5 mM, 18 h)[a]	51.6 ± 3.7
Carnosine (1 mM, 18 h)[a]	66.4 ± 1.2
Carnosine (5 mM, 18 h)[a]	74.5 ± 3.8
Carnosine (10 mM, 18 h)[a]	88.1 ± 3.1

Reproduced from Ref. [4], with permission from The Royal Society of Chemistry
[a]Exposed to 5 μM Aβ42 for 24 h afterwards

cell survival rates under the various conditions investigated. The results show that a higher concentration of carnosine induced a higher viability of PC12 cells. The results also reveal that in the control experiments, PBS buffer and carnosine alone had no effect on the cell survival.

3.4 Conclusions

In this chapter, based on the established analysis platform combined microfluidic devices with ESI-Q-TOF MS, cell secretions were collected, purified, identified, and analyzed. The carnosine protection effect against Aβ42-induced glutamate release in PC12 cells was measured in order to screen the medicine for AD protection. The key technology was the integration of an MEC into the apparatus in order to remove salts and protein interference from samples obtained in biological environments, which is necessary for ESI-MS detection. This combined analysis platform allowed for parallel cell culturing and drug exposure, the pretreatment of secretion collection through MEC, and then qualitative and semi-quantitative detection by MS.

The advantages of the developed cell secretion analysis platform include the decreased consumption of chemical reagents, a highly sensitive and selective detection approach, and the potential for short experimental times at a low cost. Our investigation also demonstrated the feasibility of this cell analysis platform to analyze other cellular secretions by changing the assay conditions, and also the possibility of application for medical screening tests and for the diagnosis of a specific disease.

In Chap. 4, the two separated chips were integrated by a valve design. The valves was set in between the two functional parts of the chip to control the flow direction, in order to separate the flow between the cell culture chip and the MEC to prevent contamination, and direct the desired reagents from branch channels.

References

1. Doble A (1999) The role of excitotoxicity in neurodegenerative disease: Implications for therapy. Pharmacol Ther 81:163–221
2. Itle LJ, Pishko MV (2005) Multiphenotypic whole-cell sensor for viability screening. Anal Chem 77:7887–7893
3. Hung PJ, Lee PJ, Sabounchi P, Aghdam N, Lin R, Lee LP (2005) A novel high aspect ratio microfluidic design to provide a stable and uniform microenvironment for cell growth in a high throughput mammalian cell culture array. Lab Chip 5:44–48
4. Wei H, Li H, Gao D, Lin JM (2010) Multi-channel microfluidic devices combined with electrospray ionization quadrupole time-of-flight mass spectrometry applied to the monitoring of glutamate release from neuronal cells. Analyst 135:2043–2050
5. Shinohara H, Wang FF, Hossain SMZ (2008) A convenient, high-throughput method for enzyme-luminescence detection of dopamine released from PC12 cells. Nat Protoc 3:1639–1644
6. Chen Z, Fu QL, Dai HB, Hu WW, Fan YY, Shen Y, Zhang WP (2008) Carnosine protects against Aß42-induced neurotoxicity in differentiated rat PC12 cells. Cell Mol Neurobiol 28:307–316

Chapter 4
Microfluidic Device with Integrated Porous Membrane for Cell Sorting and Separation

4.1 Introduction

In the real biological samples detection, the cells are mostly not from one single spice. In order to obtain the single spices for testing, and get the accurate biological, the requirement for sorting and separation different types of cells from complex samples needs to be met. Several approaches were developed to make miniaturized particle-sorting devices on a microfluidic platform. Dielectrophoretic forces [1], optical tweezing forces [2], hydrodynamic/hydrophoretic forces [3], magnetic forces [4], shear-induced lift forces [5], and gravity-driven forces [6] are introduced for cell sorting. However, many of these approaches are complicated, expensive, and require additional steps to label the particles to be sorted. Furthermore, the sorting efficiency of some of these methods is insufficient for diagnostic and therapeutic applications, such as polymerase chain reaction (PCR) [7, 8] or early cancer detection by circulating tumor cells (CTCs) [9].

The cell separation based on the size differences has also been reported, in which the microscal filters were fabricated in the microfluidic devices. They mainly belong to four types: weir-type [10], pillar-type [11], cross-flow [12], and membrane filters [13]. It's a simple and direct method to sort cells through the membrane-filter, which is also easy to be integrated into a miniatrized instrument for the rapid analysis of microliter volumes, such as blood in point-of-care diagnostic tests. However, the reported sorting approaches based on microfilters for size-based separations face the disadvantages of clogging and a relatively low sorting efficiency that may be insufficient depending on the desired use.

In this work, an alternative method for particle sorting was developed, which used poly(dimethylsiloxane)(PDMS) as a porous membrane to create a filter system within a monolithic microfluidic device that we believe remedies these problems. The SU-8 posts with controllable, variable diameters are fabricated on a silicon wafer for preparing porous PDMS membranes. A thin film of PDMS is spin-coated on the wafer such that its thickness is less than the height of the SU-8 posts. The resulting thin porous PDMS film is peeled cleanly off the wafer using a specially

H. Wei, *Studying Cell Metabolism and Cell Interactions Using Microfluidic Devices Coupled with Mass Spectrometry*, Springer Theses, DOI 10.1007/978-3-642-32359-1_4,
© Springer-Verlag Berlin Heidelberg 2013

designed cured PDMS frame. In this work, the integration of a porous PDMS membrane can be achieved without resorting to plasma oxidation, which can clog microchannels. In contrast to commercially available porous membranes which only have one fixed pore size and random pattern, this method allows multiple pore sizes on a single membrane, as well as allows the device to produce a series of size fractions at the desired positions. The smallest pore size on a single membrane is limited by the resolution of the transparent mask, while even smaller pore sizes are achieved by overlapping two membranes, as described below. The white blood cells were separated from whole blood with a 99.7% separation efficiency with a microfluidic device integrated one layer porous PDMS membrane. This approach is simple, low consumption, and power-saving, which reveals a great prospects in medical care, especially for being integrated into point-of-care diagnostics.

4.2 Experimental Section

4.2.1 Reagents and Materials

SU-8 negative photoresist was obtained from Microchem (Sunnyvale, CA, USA). Poly(dimethylsiloxane) (PDMS) RTV 615 was purchased from GE Silicones (Waterford, NY, USA). Methyltrichlorosilane (MTS) was purchased from Sigma-Aldrich (St Louis, MO, USA). Bovine serum albumin (BSA) and phosphate-buffered saline (PBS) were purchased from Invitrogen Corp. (Carlsbad, CA, USA), and tween-20 was obtained from Sigma-Aldrich (St. Louis, MO, USA). Polystyrene (PS) latex microspheres with diameters of 2 and 15 μm were manufactured by Invitrogen Corp. (Carlsbad, CA, USA); and PS 20 μm microspheres were purchased from Polysciences Inc. (Warrington, PA, USA). SPR 220–7 positive photoresist, hexamethyldisilazane (HMDS), and photoresist-developing reagents SU-8 Developer and MF-26A solvent are common chemicals supplied by the Stanford Nanofabrication Facility.

Whole mouse blood, which was obtained from just-born female white mice, and human leukemia (REH) cells were donored by the Stanford School of Medicine.

4.2.2 The Integrated Cell Sorter Device

The key technique of the cell sorter is fabricating a porous membrane by using the photoresist posts on a wafer. This idea was based on previous work in the Zare laboratory [14], which was improved by using photoresist posts with greatly decreased diameters and a reusable mold. Many commercial porous membranes can achieve sizes into the nanometer scale. However, this option allows for only one size of pore on a single membrane and the pores are arranged randomly. Here we

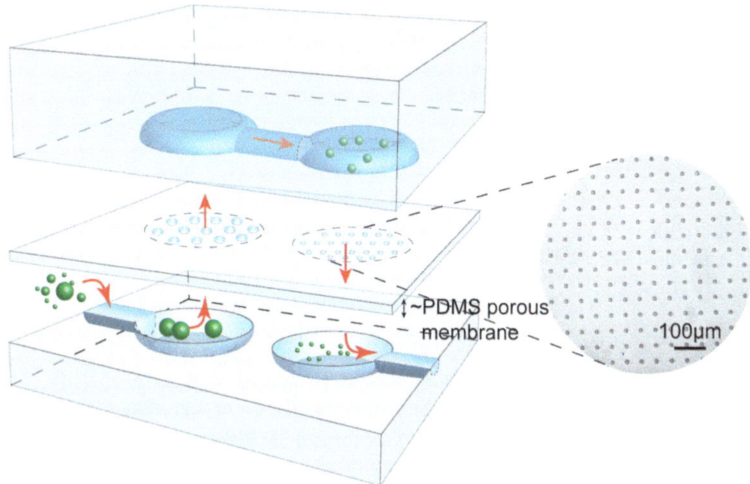

Fig. 4.1 Partial scheme of the 3D particle sorter. The *inset* shows the image of the porous membrane obtained under an optical microscope (Reproduced from Ref. [19], with permission from The Royal Society of Chemistry)

describe a porous PDMS membrane in which the pores are arranged in a specific pattern and can be designed and produced in different sizes on a single membrane. This last feature allows us to use the same membrane within a device to sort and collect different fractions of particles based on size.

Figure 4.1 shows schematics of our microfluidic device that is capable of sorting particles with three different diameters. Each filtering step is performed within an assembly of two chambers sandwiching a section of the porous membrane (Fig. 4.1, inset shows a membrane with 6.4 µm pores). The chambers have a rounded cross section because their molds were fabricated from positive photoresist that was reformed via post-development baking, which is explained in more detail below. Our integrated particle-sorting device contains two such assemblies. The analyte stream first flows from the bottom chamber to the top chamber in the first assembly, which contains the section of the membrane with the larger pores. It removes the largest particles from the stream. The stream then moves to the second assembly, where it flows from a second top chamber to another bottom chamber through the section of the membrane with the smaller pores, thus separating the particles of intermediate and smallest size from each other.

4.2.3   Fabrication of Soft Lithography Molds

Each PDMS layer was individually fabricated using a specific mold, which in turn was produced on a silicon wafer via standard photolithography techniques in the

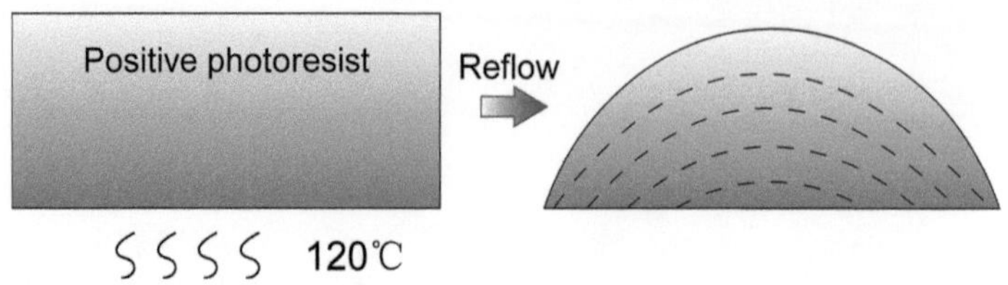

Fig. 4.2 Illustration of the sectional view of the positive photoresist, which changed from rectangular to round under the heat

Stanford Nanofabrication Facility. The patterns for this architecture were printed on transparency masks at a resolution of 40,640 dpi (dots per inch) from Fineline Imaging (Colorado Springs, CO, USA). In this work, there were two types of silicon wafer molds; the mold for the porous membrane (POM-mold) and the molds for the flow layers. With regard to the POM-mold, it should be noted that in order to form complete pores, the height of the SU-8 posts must be greater than the desired thickness of the PDMS porous membrane.

To fabricate the POM-mold, SU-8 2015 photoresist was spin-coated on an HMDS-primed silicon wafer at a thickness of 20 μm and baked at 95°C for 4 min. The photoresist on the silicon wafer was exposed under UV light through a transparency mask and baked at 95°C for 5 min, and developed with SU-8 Developer. In order to fabricate the flow layer molds, SPR 220–7 photoresist was spin-coated on an HMDS-primed silicon wafer at a thickness of 20 and 30 μm respectively for the top and bottom flow layers, then baked at 95°C for 200 s. The photoresist on the silicon wafer was exposed under UV light through a transparency mask and developed in MF-26A solvent. The flow-layer molds were baked on a hot plate (120°C for 8 min) to give the photoresist architecture a more rounded cross section, which facilitates more complete valving of the resulting PDMS channels [15] (as shown in Fig. 4.2).

Prior to be used in soft lithography, all molds were exposed to MTS vapor in a vacuum desiccator in order to prevent adhesion between the cured PDMS and the mold. The flow layer molds were exposed to MTS for 30 min, while the POM-mold was exposed for 4 h to achieve a more hydrophobic surface around the posts.

4.2.4 Fabrication of the 3D Microfluidic Device

The microfluidic particle sorter was assembled in-house in the Zare lab from three sections: the top flow layer, the membrane layer, and the bottom flow layer. The top and bottom flow layers were both fabricated as PDMS slabs with thicknesses of 7 and 2 mm, respectively, while the membrane layer was a thin PDMS film with a thickness of less than 20 μm. An imbalanced cross-linking ratio bonding method [16] was adopted in our experiments because successful alignment of the layers required a low error tolerance, and this method allowed us to continually correct alignment. The principle of imbalanced cross-linking ratio bonding is based on the theory that the cross-agent can migrate through the interface between cross-linker-poor and cross-linker-rich layers. All layers linking to be aligned are cured into a solid state that can easily be removed from the molds. Despite their cured nature, the layers retain a sufficient imbalance with regard to their cross-linked state such that an effective bond can form between them at their interface. Thus, the imbalanced cross-linking ratio bonding method allows for multiple alignment attempts, because both layers are solid, without damaging either layer.

The fabrication process is described as follows:

1. Well-mixed liquid PDMS prepolymer was poured into the molds of the top and bottom flow layers. The prepolymer for the top flow layer was mixed with an excess of cross-linker (mass ratio 5:1 RTV A:B), while the prepolymer for the bottom layer was mixed with a deficiency (mass ratio 20:1 RTV A:B). After degassing both layers under a vacuum, they were cured in an 80°C oven for 1 h.
2. The cured PDMS slabs (the top layer was 7 mm thick and the bottom layer was 2 mm thick) were cut with a scalpel and peeled from the mold. Holes were punched through the channel inlets of the top flow layer with a syringe needle that had the tip sawed off.
3. A 10:1 mass ratio of PDMS was diluted in a one-half mass equivalent of cyclohexane. The mixture was spin-coated on the POM-mold with an initial spin rate of 500 rpm for 18 s and a final spin rate of 3,000 rpm for 60 s. The mold was placed on a flat surface at room temperature for 40 min to allow the surface of the PDMS to smoothen, and then inserted into an 80°C oven for 20 min to cure the membrane layer.
4. The top flow layer prepared in Step 2 was placed on the PDMS-coated POM-mold and aligned by inspection through a stereoscope. Afterwards, a 5:1 PDMS prepolymer mixture was poured around the aligned layers and then cured at 80°C for 1 h.
5. The chips were cut from the POM-mold. Access holes were punched through the combined layers at positions corresponding to the inlets and outlets of the bottom flow layer.
6. The resulting chips in Step 5 were placed on the bottom flow layer prepared in Step 1, aligned by inspection, and sealed with gentle pressure. The whole device was placed in an 80°C oven overnight for final bonding.

4.2.5 Samples and Setup

Whole mouse blood was collected from white mice into Microtainer tubes containing dipotassium EDTA (Fisher Scientific, Montreal, QC, USA). Blood samples were used within 1 day of collection. The REH human leukemia cell lines were grown in RPMI (Roswell park memorial institute) 1640 medium (Invitrogen) supplemented with 10% fetal bovine serum, 25 mM HEPES and L-glutamine in an incubator were

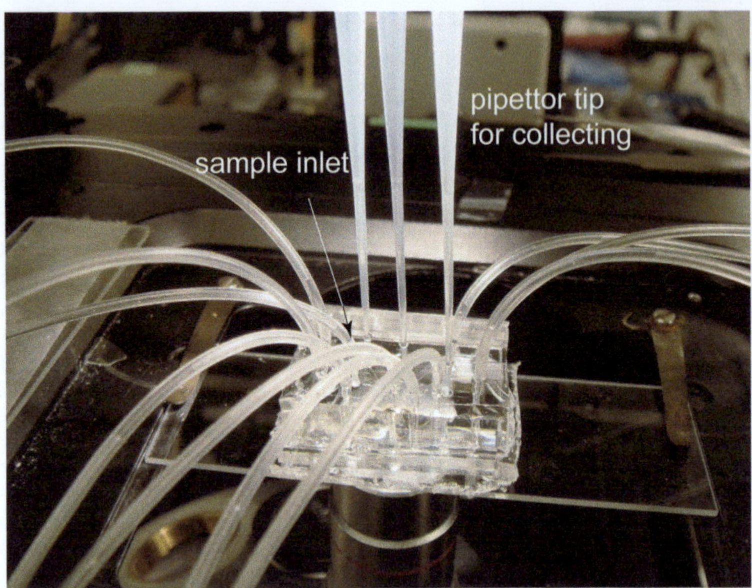

Fig. 4.3 The experimental setup when the cell sorting experiments were carried out on the microfluidic device with porous PDMS membrane

maintained at 37°C with 5% CO_2. The REH cells used in the experiment were obtained after 3–4 days of culturing, which had a rough concentration of $10^6/\mu L$. The whole blood sample was diluted by a factor of 40 in PBS buffer. REH cell suspension was mixed with diluted whole blood in a 1:1 ratio before injection into the particle sorter.

The CellTracker Orange CMTMR (5-(and-6)-(((4-chloromethyl)benzoyl)amino) tetramethylrhodamine) was applied in the experiments to reveal the activity of the cells, and to distinguish the REH cells from RBCs. The CellTracker was prepared in the serum-free culture medium as 5 μM under the product instruction. The warm CellTracker solution (37°C) was added into the cells which are centifugated without supernatant. Afterwards the cells were incubated at 37°C for 45 min. The supernatant containing CellTracker was removed after centifugation. The cells was gently resuspended in warm (37°C) medium, and indubated at 37°C for 30 min. Finally the labeled cells were washed by PBS buffer, then mixed with whole blood cells for the following experiments.

Integrated microfluidic devices were filled and flushed with 0.2% (v/v) aqueous solution of tween-20 before loading the PS beads and 1% (w/v) BSA solution in PBS buffer before loading blood samples. Sample mixtures were introduced from the inlet, and moved along the channels using a homemade pressure controller in the Zare lab (as shown in Fig. 4.3). A microscope (Nikon ECLIPSE TE2000-U, Kanagawa, Japan) equipped with a CCD camera (Mintron MTV-63KR11N,

Fremont, CA, USA) was used to observe and obtain images of the microfluidic device. Upon completion of separation, the samples were collected at the different outlets by gel-loading pipet tips, diluted with the Isoton diluent (Beckman Coulter Inc., Brea, CA, USA), and further analyzed for enumeration and size distribution measurements using a Coulter counter (Model Z2, Beckman Coulter Inc.) located in the Stanford School of Medicine (Stanford, CA, USA).

4.3 Results and Discussion

4.3.1 Fabrication of PDMS Porous Membrane

As discussed, the porous membrane is an essential component of the particle sorter, serving as a filter to separate particles by their sizes. The main requirement for forming the porous PDMS membrane on the POM-mold is that the thickness of the membrane should be smaller than the height of the photoresist posts in order to ensure completely through pores. However, the membrane must also be thick enough to withstand peeling from the membrane as well as the fluidic pressure generated during the chip's use. In order to achieve a sufficiently thin membrane, the PDMS prepolymer was diluted in cyclohexane to decrease the viscosity. The cyclohexane was expected to have evaporated quickly during the curing step in an 80°C oven. However, the dilution of the PDMS prepolymer negatively impacts the toughness of the membrane.

Figure 4.4 presents a plot of the thickness of the PDMS membrane versus the dilution factor of PDMS prepolymer in cyclohexane as well as the spin-coating speed. First, PDMS prepolymer in varying dilutions of cyclohexane was spin-coated at 500 rpm for 18 s followed by 3,000 rpm for 60 s. The data show that a mixture with a dilution factor of 0.2 or less generated membranes with thicknesses larger than 20 μm, which is the height of the photoresist posts on the POM-mold. Considering the toughness and 20 μm limit of the membrane, the thickest membrane under 20 μm, which was obtained from a dilution factor of 0.5, was adopted. Further experiments determined the optimum spin-coating speed under a dilution factor of 0.5. The data show that at a spin-coating speed higher than 2,600 rpm, the resulting thickness of the PDMS membrane is less than 20 μm. Again considering toughness and the 20 μm limit, the spin-coating condition for fabricating all the porous PDMS membranes was set to 500 rpm for 18 s followed by 3,000 rpm for 60 s. To summarize, these experiments led us to adopt a dilution factor of 0.5 and a spin-coating speed of 3,000 rpm to obtain the thickest possible membrane under 20 μm.

To make the porous membrane, the PDMS prepolymer mixture was spin-coated on the POM-mold according to the above settings and cured in an oven at 80°C for 20 min. To peel off the porous membrane without damaging either the membrane or the photoresist posts, specially designed PDMS structures were used, the requisite

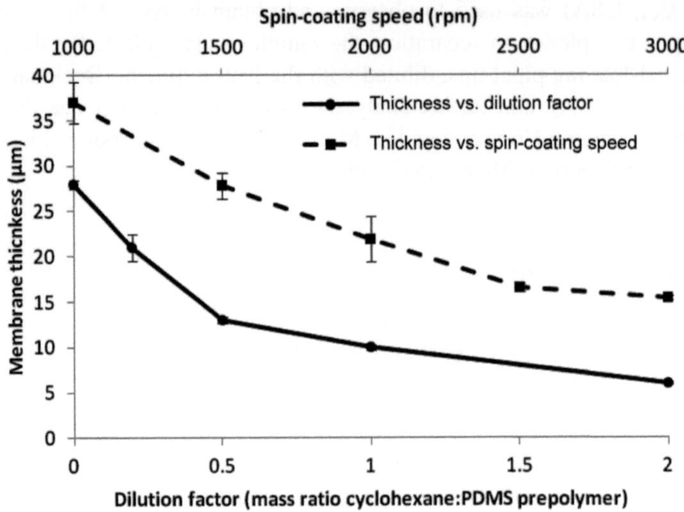

Fig. 4.4 Plot of the thickness of spin-coated PDMS membranes at various cyclohexane dilutions and spin-coating speeds. The *solid line* represents membrane thickness versus dilution factor of PDMS prepolymer in cyclohexane, spun at 500 rpm for 18 s followed by 3,000 rpm for 60 s. The *dashed line* represents thickness versus the spin-coating speed, using a dilution factor of 0.5 and a coverage spin speed of 500 rpm for 18 s

height and width of which were determined according to the dimensions of the region containing the photoresist posts. Two alternative approaches were used to peel off the porous membrane from the POM-mold (shown in Fig. 4.5).

First, a specifically fabricated PDMS support frame was used to bond with the membrane outside of the area containing the photoresist posts. The PDMS frame was cut off after bonding the membrane with other layers. Second, one of the flow layers could be aligned and bonded to the porous membrane directly. The chamber on the flow layer should be sufficiently tall and wide so as not to contact the photoresist posts. By these means we were able to peel the membrane from the POM-mold without removing the photoresist posts. The microscopic photos of the porous PDMS membrane with two different opening sizes were shown in Fig. 4.6.

4.3.2 Fabrication and Pore Size Control of the POM-Mold

Although more expensive transparency or chrome masks that provide higher resolution are commercially available, a transparency mask with 40,640 dpi resolution was adopted in our experiments, which is much cheaper and more convenient for fabricating the mask for the POM-mold.

When choosing a photoresist for fabrication of the POM-mold, we considered two main requirements. First, photoresist posts with large aspect ratios were needed

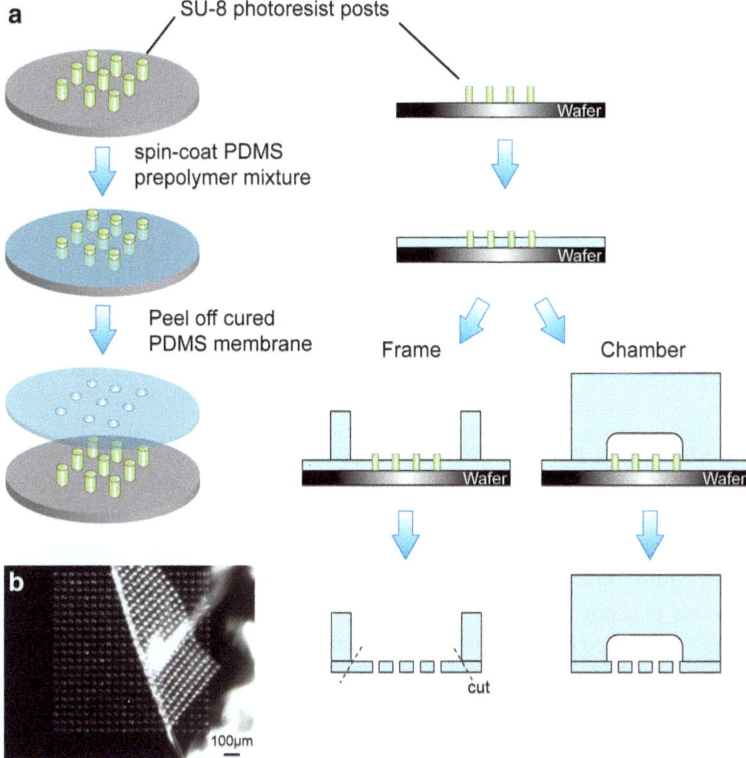

Fig. 4.5 (**a**) Schematic of the procedure for making a porous PDMS membrane from the POM-mold. Illustration of the cross section is presented on the right. (**b**) Image of the porous PDMS membrane being peeled from the POM-mold (Reproduced from Ref. [19], with permission from The Royal Society of Chemistry)

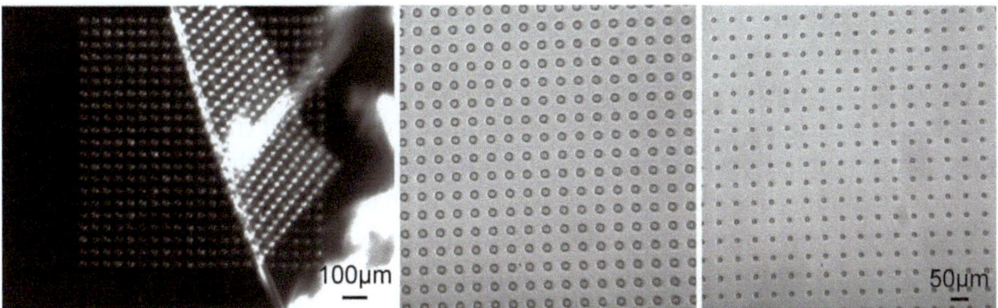

Fig. 4.6 The microscopic photos of the POM mold fabricated by SU-8 2015 and the obtained porous PDMS membranes. The three pictures from *left* to *right* are: the process of peeling the porous PDMS membrane off the POM mold; porous membrane with opening of 16.6 µm; porous membrane with opening of 6.4 µm

Table 4.1 Comparison of negative and positive photoresists for the fabrication of different pore sizes

Photoresist	Expected poresize (μm)	Printed pore size (μm) on the mask	Measured pore size (μm) on the PDMS membrane
SU-8 2015 negative	8	6.3	6.4 ± 0.3
	9	7.5	7.8 ± 0.3
	10	9.3	9.5 ± 0.4
	17	16.3	16.6 ± 0.3
SPR 220–7 positive	8	8.9	3.8 ± 1.9
	9	9.7	5.6 ± 1.6
	10	10.5	7.1 ± 0.7
	17	17.1	15.7 ± 0.4

in order to produce a thick membrane (tall photoresist posts) with small pores (narrow photoresist posts). However, there is an upper limit for the aspect ratio that a given photoresist material can achieve. Second, the photoresist posts should adhere firmly to the wafer surface so that they can survive photolithography and repeated uses in soft lithography. With these requirements in mind, we tested the abilities of two available photoresists, SU-8 2015 (a negative resist) and SPR 220–7 (a positive resist readily supplied by the Stanford Nanofabrication Facility), to form the desired structures 20 μm in height. To enhance the adherence of the photoresist to the wafer surface, silicon wafers were cleaned with standard piranha solution and treated with HMDS prior to spin-coating. We determined that SPR 220–7 demonstrates superior surface adhesion compared to SU-8 2015. However, SPR 220–7 was not capable of forming a 20 μm thick layer in a single spin-coating run, which meant that two coats were required.

Table 4.1 presents a comparison of expected pore sizes, their printed sizes on the masks, and the actual pore sizes obtained on the PDMS porous membranes. We printed masks for positive and negative photolithography with four expected pore sizes: 8, 9, 10, and 17 μm. It is important to note that our masks had a limited resolution that impacted the size of the printed images. For the negative mask used with SU-8 2015, where the pore image was transparent and the mask background was black, pore images were always smaller than the expected sizes. Conversely, the pore images on the positive mask required by SPR 220–7, with black pore images printed on a transparent background, were larger than the expected sizes.

Following photolithography and subsequent fabrication of porous PDMS membranes, we determined that the size distribution of pores generated from molds composed of negative photoresist appeared to be more regular than that of pores generated from positive photoresist molds. The reasons for this finding are: (1) due to the different light sensitivities, a much longer exposure time is required for the positive photoresist, which induces diffraction and over-exposure. As the result, the sizes of the photoresist posts and the pores that are made from them are reduced, which turn out uncontrollable size of the photoresist posts patterned on the mold; and (2) the low viscosity of positive photoresist meant that two coats had

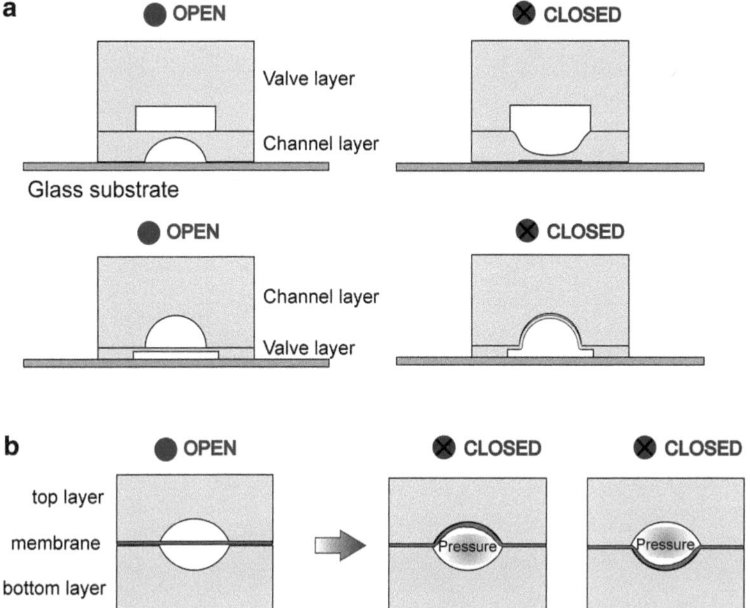

Fig. 4.7 Microvalve designs which are integrated in microfluidic devices. (**a**) Two microvalve designs normally used. (**b**) Three-state valve design used in this work

to be applied to obtain a 20 μm-thick layer, which results in a rougher surface. Additionally, the data in Table 4.1 show that, for positive photoresist, the magnitude of the error increases for smaller pores. Briefly, the character of positive photoresist determined the limit of usage for the fabrication of smaller posts. Therefore when our experiment called for pores with a regular controllable size under 15 μm, negative photoresist SU-8 2015 was selected to fabricate the POM-mold. However, in order to obtain pores larger than 15 μm, positive photoresist SPR 220–7 was an alternative option because it exhibited stronger adherence to the surface substrate.

4.3.3   Microvalve System for Cell Sorting

In normal microfluidic device integrated with microvalve structure, the flow channels have smooth curved surfaces derived from reshaped photoresist, while the valve channels are cast from rectangular photoresist. Normally, flow and valve architectures are maintained on separate layers, with the two main styles showed in Fig. 4.7a.

However, in our work we required a 3D microfluidic structure in which a porous membrane is sandwiched between two flow layers, so that a much longer interconnection between the valves and their inlets would not be avoided when the

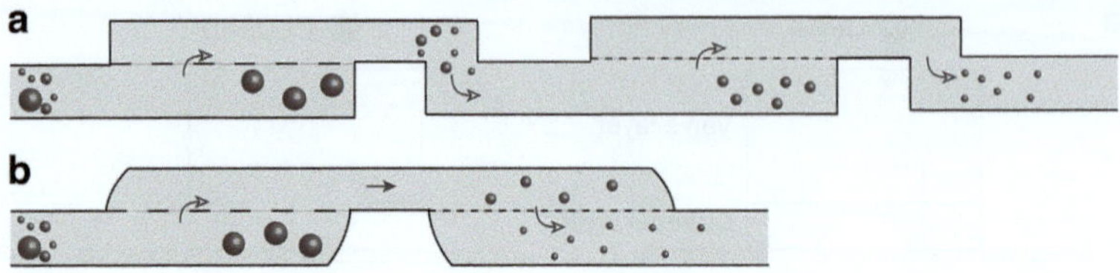

Fig. 4.8 Scheme of the sectional view of the particle sorter. (**a**) Normal valve design integrated. (**b**) Three-state valve design integrated

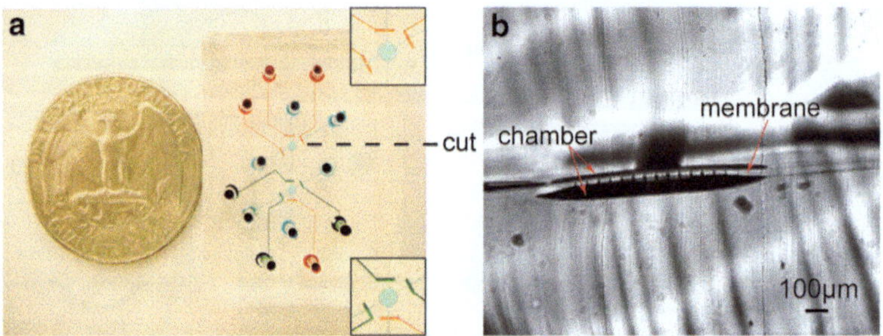

Fig. 4.9 (**a**) Picture of the three-size particle sorter. The *insets* show enlarged images of the two chamber areas. (**b**) Image of the cross-section along *dashed line* of (**a**) (Reproduced from Ref. [19], with permission from The Royal Society of Chemistry)

normal valve designmethodology was adopted (Fig. 4.8a). A longer interconnection would introduce more resistance and thus require higher pressure to close the valves, which would cause more problems during chip operation. To solve this problem, we developed a three-state valve (Fig. 4.7b) to reduce the interconnection inside the structure. In this case, flow channels and valve channels were fabricated on the same positive resist mold, allowing for a simpler design that incorporates shorter interconnection and gives both flow and valve architecture a rounded cross section. The porous PDMS membrane sandwiched between the two layers also acts as the thin film for closing a flow channel as a valve. The membrane is not porous in these areas.

Figure 4.9 shows a picture of a fabricated particle sorter. Channels filled with blue dye represent the flow channels, interconnected across layers to form a 3D structure. Channels filled with red and green dyes are valves which, as stated above, are fabricated in the top and bottom flow layers, respectively. A scalpel was used to cut through the chamber along the dashed line, and the cross section was examined under an optical microscope. In Fig. 4.9b, the through-pores and the PDMS membrane layer can be clearly seen between the two flow layers. The streaks on the images are caused by mechanical cutting from the scalpel.

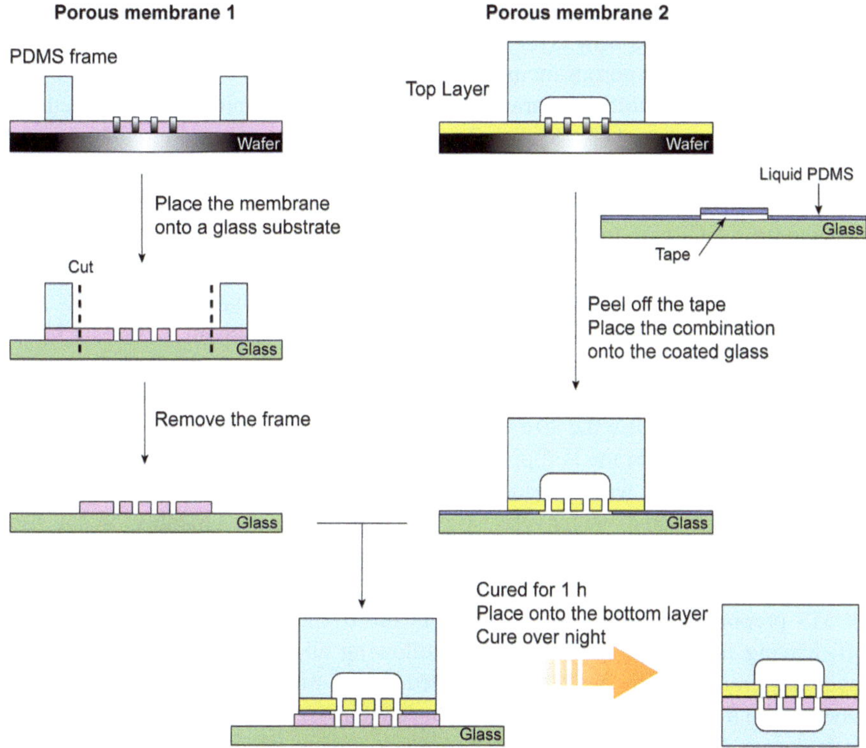

Fig. 4.10 Schematic of the fabrication procedure for the overlapped aligned porous membrane. Different PDMS membranes are labeled with different *colours* (Reproduced from Ref. [19], with permission from The Royal Society of Chemistry)

4.3.4 Aligned Overlap of Porous Membranes for the Fabrication of Smaller Pores

It is possible to create even smaller pores within a PDMS membrane than those described above by aligning two porous membranes such that their pores overlap. The size of the overlap determines the size of the new, smaller pore. This was accomplished in the following manner (Fig. 4.10). Membrane 1 was formed from a PDMS prepolymer, mixed in 5:1 mass ratio and then diluted in cyclohexane by a factor of 2. Because the viscosities of PDMS RTV A and B are similar, this ratio adjustment did not cause any apparent change to the membrane thickness from the results obtained previously. Following curing at 80°C, a PDMS frame was used to peel off porous membrane 1 from the POM-mold. The membrane was carefully placed onto a pre- cleaned glass slide with care to prevent trapping air between the membrane and the glass slide. Due to electrostatic interaction, the thin film tightly attached to the glass surface. A scalpel was used to remove the PDMS frame and the redundant film on the edges of porous membrane 1. Meanwhile, porous membrane 2 was bound to the top flow layer according to the procedures described above for fabrication of a single porous membrane.

The two porous PDMS membranes were bound by the dipping-attaching method, as described elsewhere [17, 18]. Briefly, PDMS prepolymer (mixed in a 10:1 mass ratio), diluted in cyclohexane by a factor of 3, was spin-coated on a glass slide at 3,300 rpm to form a thin film. To form an even thinner film, the glass slide was put into an 80°C oven to cure for 2 min before it was used. A PDMS layer was then dipped into the mortar and attached to another PDMS layer. Thus, between the two PDMS layers a portion of the thin film of PDMS prepolymer was present, which was then cured in an oven at 80°C for 1 h to become an adhesive PDMS layer.

However, for the porous PDMS membrane, the normal dipping-attaching method is not applicable because the size of the pores is so small that the mortar clogs the pores when the membrane is dipped into it. To prevent pore clogging, a partially "mortar-free" method was introduced for bonding two porous PDMS membranes. A small piece of tape, which was slightly larger than the chamber on the top flow layer, was pasted onto a specific position of the glass slide, corresponding to the position of the chamber on the top flow layer. After spin-coating the PDMS prepolymer solution, and curing in the oven, the cover tape was peeled off, leaving that space free of mortar. Following alignment of the top chamber layer to porous membrane 2, the combined PDMS structure was dipped into the mortar such that the mortar-free area was located under the chamber. The combined structure was then aligned and bound to porous membrane 1 prior to curing it in an 80°C oven for 1 h. The whole device was finished by bonding the resulting three layers with the bottom flow layer by curing the entire chip in an 80°C oven overnight.

Figure 4.11 shows a schematic illustration and microscopic images of the aligned, overlapped porous membrane. Taking the overlapped pores in Fig. 4.11b as an example, an overlapping area with a diameter ranging between 2.5 and 3.3 μm was created. Additionally, three-layer overlapped pores were achieved according to the same procedures, as shown in Fig. 4.11d.

Due to the resolution limitations of the mask, no sharply angled shapes could be fabricated for the pores on the membrane. The production of triangular- and rectangular-shaped pores exhibited rounded corners instead of sharp angles. The error in the size of the smaller, overlapped pore that is created as a result of this process has not yet been characterized. Regardless, we have demonstrated a novel, relatively cheap and surprisingly simple method of fabricating customizeable PDMS membranes for use in a monolithic particle-sorting device. As an aside, one may avoid cells and particles becoming trapped between the membranes via frequent 'flush-and-collect' cycles, as well as treating the surface with polyvinyl acetate in the case of cells to avoid adhesion.

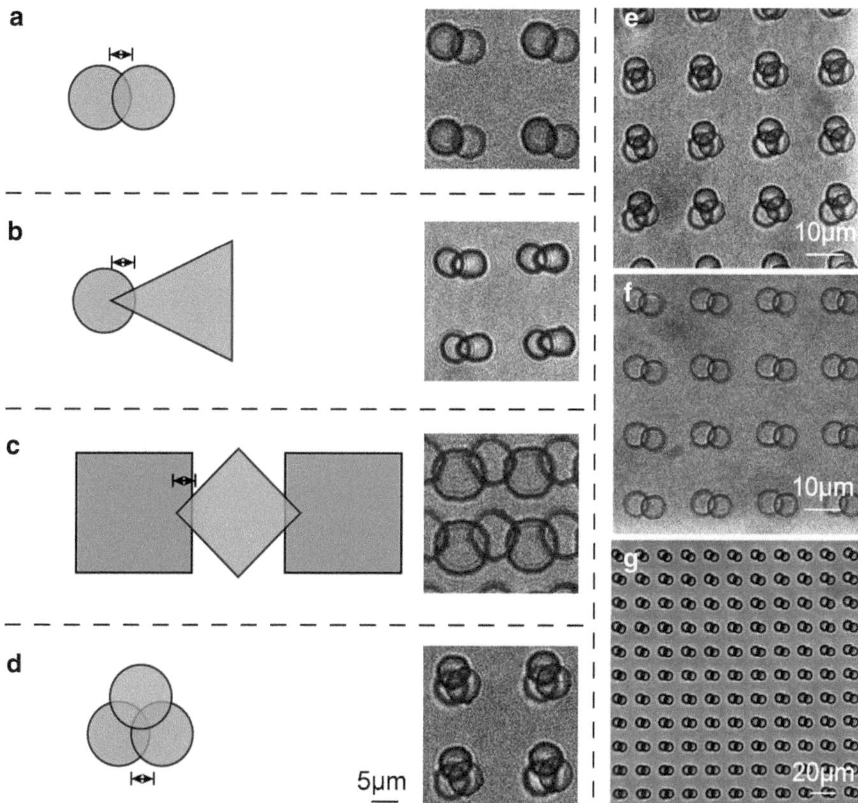

Fig. 4.11 (**a**)–(**d**) Schematic illustrations of the expected *pore shapes* are shown in the *left column*, while microscopic images of the overlapped aligned porous membranes are shown in the *right column*. (**e**) and (**f**) Microscopic images of the overlapped pore array. (**g**) Microscopic images of the overlapped pore array showing a larger area (Reproduced from Ref. [19], with permission from The Royal Society of Chemistry)

4.3.5   Polystyrene Microbeads Sorting Process

To demonstrate the effectiveness of the particle sorter, we tested the separation of polystyrene beads with different diameters as a proof-of-concept. PS beads with a diameter of 2.5, 15, and 20 μm were mixed in a 0.2% tween-20 solution for evaluation of the device's sorting efficiency. For this demonstration, a device was fabricated in which the membrane contained 16.6 μm pores in the first assembly and 6.4 μm pores in the second. Figure 4.12 illustrates the sorting procedure. The chip consists of two sections: the main flow channel (straight channel with valves 1, 4, 7 turned off) and the flushing channels (beveled channels with the valves 2, 3, 5, and 6 turned on). Valves 1, 2, 3, and 7 are on the top layer, and valves 4, 5, and 6 are on the bottom flow layer.

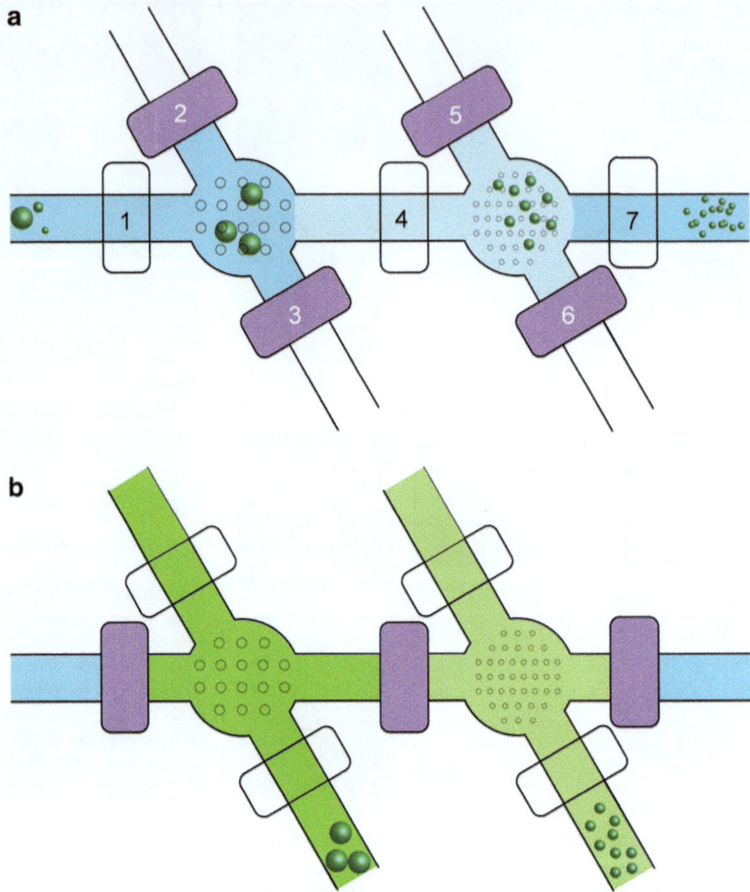

Fig. 4.12 Procedure for sorting particles with three different diameters. The *darker color* represents liquid flow in the *bottom layer*; the *lighter color* represents liquid flow in the *top flow layer* (Reproduced from Ref. [19], with permission from The Royal Society of Chemistry)

In our device, sorting particles is a two-step process. First, the particle suspension is introduced into the inlet with a pressure of 10 psi. Particles are directed though the main flow channel because valves 2, 3, 5, and 6 are closed. When the mixture meets the first chamber, the 20 μm particles are blocked by the porous membrane and gather in the bottom layer, while the 15 and 2.5 μm particles go through the pores and continue flowing in the channel on the top flow layer. The 15 μm particles are then trapped in the second chamber on the top flow layer while the 2.5 μm particles are collected at the end of the main flow channel on the bottom layer. Second, the device is reconfigured for collection of particles by closing valves 1, 4, and 7 and opening valves 2, 3, 5, and 6. A 0.2% tween-20 solution is injected into the beveled channels to flush the 15 and 20 μm particles to the outlets in order to collect them.

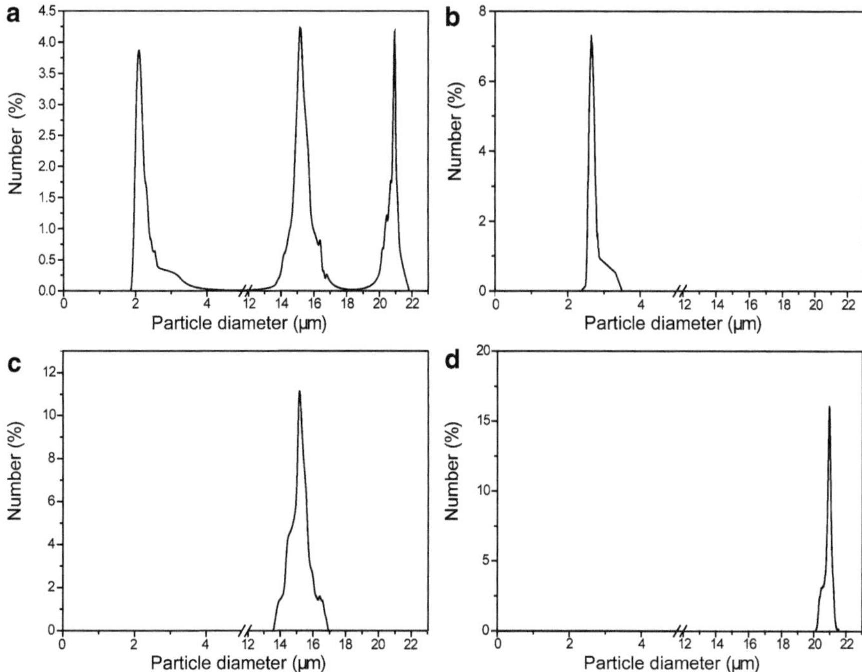

Fig. 4.13 Sorting results for different sizes of PS particles. (**a**) Particle mixture measurement before sorting; (**b**), (**c**), and (**d**) measurement of the smallest, intermediate and largest collected particles, respectively (Reproduced from Ref. [19], with permission from The Royal Society of Chemistry)

New samples can then be injected to continue sorting. To prevent the generation of air bubbles, the channels were filled with deionized water containing 0.2% tween-20 prior to the sorting process.

Clogging was successfully avoided by flushing the chamber and collecting the samples frequently. The separation efficiency of the separated sample was defined as the ratio of the number of particles with a diameter inside the target size range to the total number of particles collected. The size distribution measurements from the Coulter counter (Fig. 4.13) revealed that the separation efficiency of separated samples in each collector was greater than 99.9%, as determined from the data output from the Coulter counter. This experiment was repeated 15 times.

4.3.6 Whole Blood Cells Sorting Process

Whole blood cells were obtained to evaluate the efficiency of sorting particles within a biologically relevant sample. Please note that white blood cells are at a

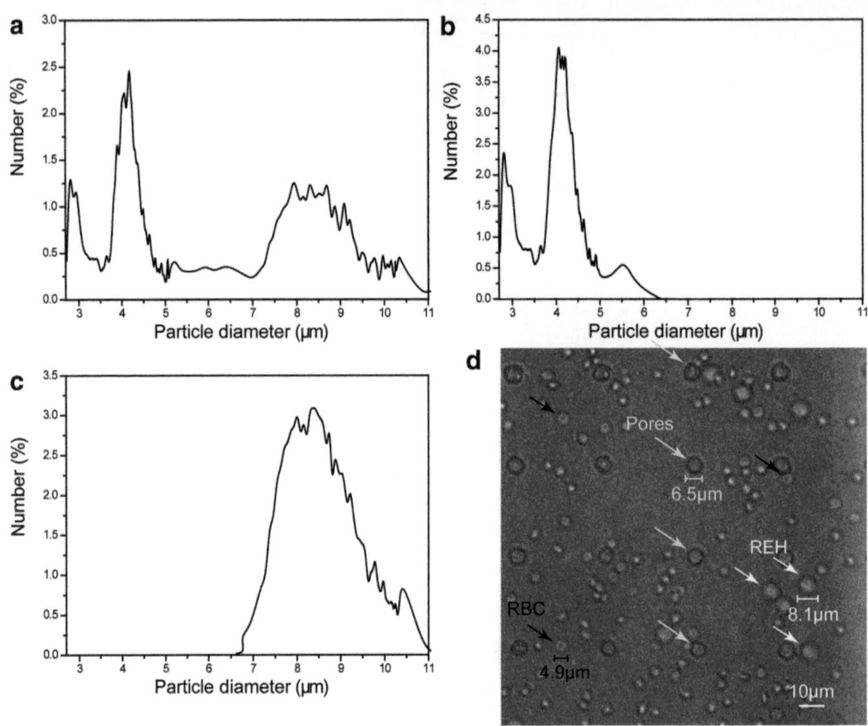

Fig. 4.14 Sorting results of whole blood. (**a**) Particle distribution measurements of whole blood before sorting; (**b**) particle distribution measurements of collected red blood cells and platelets; and (**c**) collected white blood cells and REH cells. (**d**) Microscopic images inside the sorting chamber. *Red arrows* point at red blood cells (RBC), and *yellow arrows* point to REH cells, respectively (Reproduced from Ref. [19], with permission from The Royal Society of Chemistry)

concentration 1,000th of that of red blood cells, so their number was supplemented with similar-sized REH cells so that the separation of the two differently sized groups could be more precisely evaluated.

White blood and REH cells were separated from whole blood by moving them through a porous membrane with a pore size of 6.4 μm. Because red blood cells have a disk diameter of 4–5 μm, and white blood cells and REH cells have a diameter of 7–10 μm (Fig. 4.14), the white blood and REH cells were blocked while the red blood cells and platelets went through the pores to be collected at the outlet. For each separation experiment, 10 μL samples were loaded. Approximately 10^7 cells were sorted in each separation that had the duration of 3–5 min. Figure 4.14 presents the sorting results evaluated via a Coulter counter. The mean diameter was 5.982 μm, and ~40% of the cells had a diameter larger than 6.4 μm. Following separation, size measurements of the purified sample that contained cells with a diamater less than 6.4 μm had a separation efficiency greater than 99.9% and a mean diameter of 4.056 μm (Fig. 4.14b). The purified sample that contained cells with a diameter greater than 6.4 μm was found to have a separation efficiency of 99.7% and a mean diameter of 8.688 μm (Fig. 4.14c).

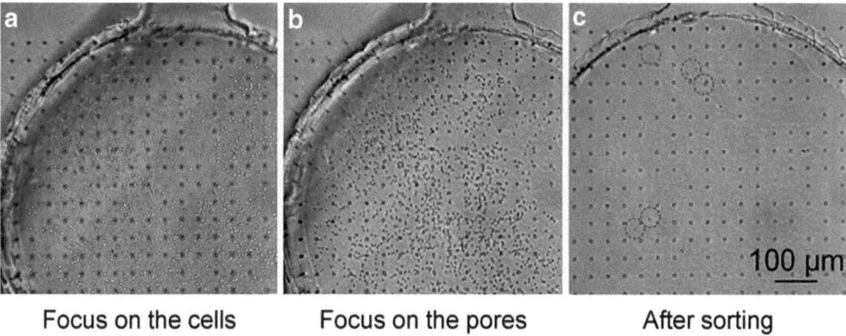

Fig. 4.15 Images of the same chamber during and after a sorting experiment with whole blood are shown. (**a**) The chamber filled with sample, with the camera focused on the cells. (**b**) The same as (**a**), but focused on the pores. (**c**) The chamber after sorting a sort and collection cycle without an extra flushing step. *Grey dotted circles* mark the few REH cells which adhered to the membrane. These were removed with extra flushing step

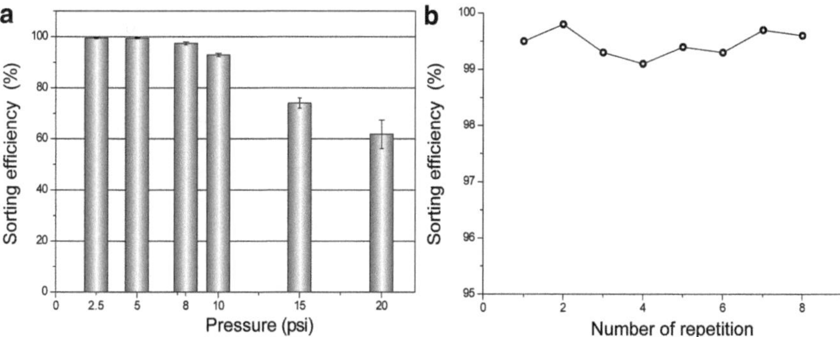

Fig. 4.16 (**a**) Plot of sorting efficiency versus driving pressure. A 99.7% sorting efficiency could be achieved at or below a pressure of 5 psi. A poor efficiency was observed for pressures at or above 15 psi. (**b**) The whole blood sorting experiment was performed eight times experiments and demonstrated good reproducibility with an average sorting efficiency of 99.7%

The whole device was treated by 1% (w/v) BSA solution in PBS before loading the blood samples to avoid adhereing to the channel wall. Clogging was prevented by flushing the filter chamber after every separation and collection cycle (Fig. 4.15).

While a stronger driving pressure will produce a faster separation, it will also result in a poorer sorting efficiency caused by the increased flexibility of cells compared to the polystyrene beads. We evaluated and optimized the influence of increased driving pressure on the sorting efficiency. Figure 4.16a shows this relationship. The data show that with a pushing pressure of less than 5 psi, 99.7% sorting efficiency was achieved. When the driving pressure was increased to 8 psi, the sorting efficiency decreased to 97.4%. The drop in efficiency became more dramatic as the driving pressure was increased beyond 10 psi. In order to maximize

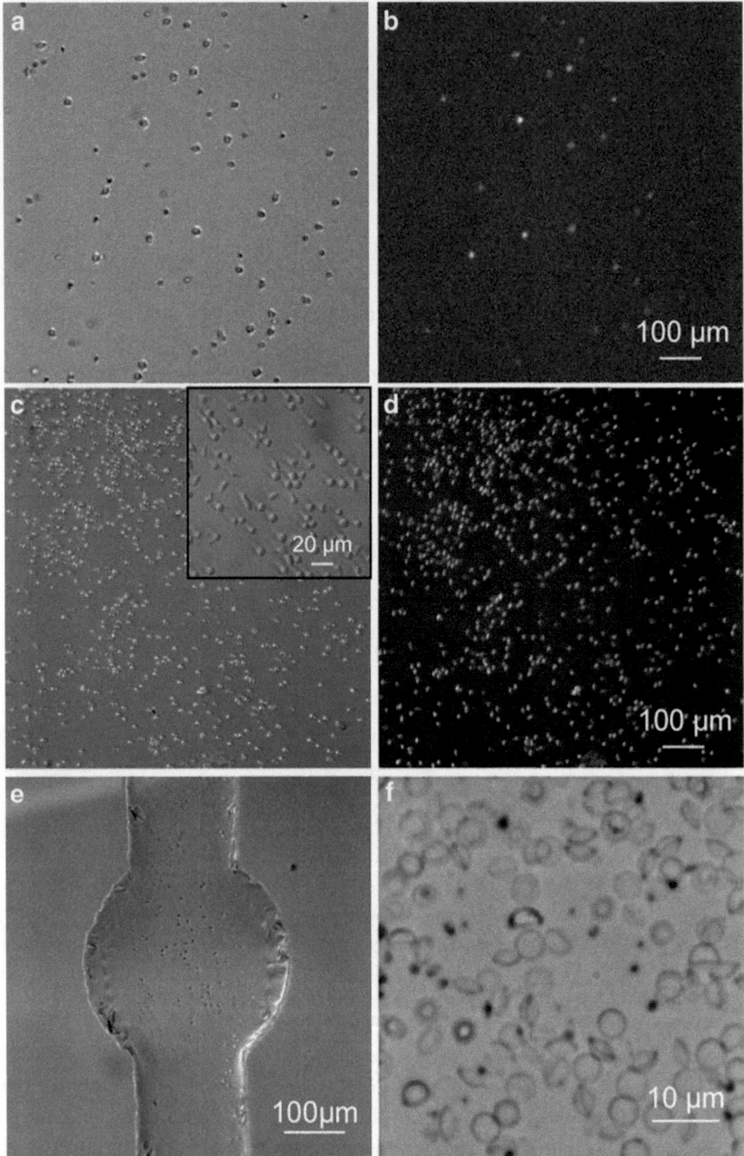

Fig. 4.17 Images of cell samples before sorting and after collection (Reproduced from Ref. [19], with permission from The Royal Society of Chemistry)

sorting speed without significantly sacrificing sorting efficiency, a driving pressure of 5 psi is suggested. The reproducibility was evaluated under driving pressure of 5 psi (Fig. 4.16b).

The viability of cells was qualitatively confirmed to remain high following the sorting process (Fig. 4.17). REH cells were premarked with CellTracker Orange

CMTMR, which can visualize the viability of cells. In Fig. 4.17b, the microscopic photos of the original sample containing mouse whole blood and marked REH cells were showed, under bright field and fluorescence respectively. The marked REH cells can be observed. Figure 4.17d showed the photos of the collected REH cells and WBCs from the whole blood sample, under bright field and fluorescence respectively. The inset on Fig. 4.17c shows the enlarged image of the cells. The fluorescence intensity confirmed the viability of the REH cells was still high after being sorted from the whole blood sample. Figure 4.17 showed the image and an enlarged image of the channel that collects RBCs and platelets. It can be observed that the RBCs remained well pancake shape, which revealed the sorting throught the porous PDMS membrane didn't reduce the cells viability.

4.4 Conclusions

In this chapter, we have described a microfluidic particle sorter which incorporates a PDMS porous membrane. Membranes containing pores as small as 6.4 μm were easily fabricated on a POM-mold, and even smaller pores were generated by aligning two or more membranes so that the pores overlapped. A sorting efficiency greater than 99.9% for polystyrene beads and a 99.7% sorting efficiency for whole blood were achieved from our particle sorter. We were able to separate white blood cells, as well as red blood cells and platelets combined, from whole blood.

Additionally, the size of the pores could be conveniently changed to fit the specific objective, such as CTCs. We believe that these results demonstrate the potential of this simple monolithic microfluidic particle sorter to be integrated in miniature and automatic equipment for point-of-care diagnostics. Further work is needed to perfect this setup for isolation, collection, and enumeration of blood components, but the present results clearly demonstrate the principles of this new separation technique.

References

1. Vahey MD, Voldman J (2008) An equilibrium method for continuous-flow cell sorting using dielectrophoresis. Anal Chem 80:3135–3143
2. Lin CH, Lin CC, Chen A (2008) Microfluidic cell counter/sorter utilizing multiple particle tracing technique and optically switching approach. Biomed Microdevices 10:55–63
3. SooHoo JR, Walker GM (2009) Microfluidic aqueous two phase system for leukocyte concentration from whole blood. Biomed Microdevices 11:323–329
4. Huang R, Barber TA, Schmidt MA, Tompkins RG, Toner M, Bianchi DW, Kapur R, Flejter WL (2008) A microfluidics approach for the isolation of nucleated red blood cells (NRBCs) from the peripheral blood of pregnant women. Prenatal Diagn 28:892–899
5. Toner M, Di Carlo D, Edd JF, Irimia D, Tompkins RG (2008) Equilibrium separation and filtration of particles using differential inertial focusing. Anal Chem 80:2204–2211

6. Huh D, Bahng JH, Ling YB, Wei HH, Kripfgans OD, Fowlkes JB, Grotberg JB, Takayama S (2007) Gravity-driven microfluidic particle sorting device with hydrodynamic separation amplification. Anal Chem 79:1369–1376
7. Wilding P, Kricka LJ, Cheng J, Hvichia G, Shoffner MA, Fortina P (1998) Integrated cell isolation and polymerase chain reaction analysis using silicon microfilter chambers. Anal Biochem 257:95–100
8. Da X, Zhang C (2010) Single-molecule DNA amplification and analysis using microfluidics. Chem Rev 110:4910–4947
9. Cristofanilli M, Budd GT, Ellis MJ, Stopeck A, Matera J, Miller MC, Reuben JM, Doyle GV, Allard WJ, Terstappen LWMM, Hayes DF (2004) Circulating tumor cells, disease progression, and survival in metastatic breast cancer. N Engl J Med 351:781–791
10. Groisman A, VanDelinder V (2006) Separation of plasma from whole human blood in a continuous cross-flow in a molded microfluidic device. Anal Chem 78:3765–3771
11. Mohamed H, Turner JN, Caggana M (2007) Biochip for separating fetal cells from maternal circulation. J Chromatogr A 1162:187–192
12. Groisman A, VanDelinder V (2007) Perfusion in microfluidic cross-flow: separation of white blood cells from whole blood and exchange of medium in a continuous flow. Anal Chem 79:2023–2030
13. Yobas L, Ji HM, Samper V, Chen Y, Heng CK, Lim TM (2008) Silicon-based microfilters for whole blood cell separation. Biomed Microdevices 10:251–257
14. Luo YQ, Zare RN (2008) Perforated membrane method for fabricating three-dimensional polydimethylsiloxane microfluidic devices. Lab Chip 8:1688–1694
15. Studer V, Hang G, Pandolfi A, Ortiz M, Anderson WF, Quake SR (2004) Scaling properties of a low-actuation pressure microfluidic valve. J Appl Phys 95:393–398
16. Eddings MA, Johnson MA, Gale BK (2008) Determining the optimal PDMS-PDMS bonding technique for microfluidic devices. J Micromech Microeng 18:1–4
17. Wu HK, Huang B, Zare RN (2005) Construction of microfluidic chips using polydimethyl-siloxane for adhesive bonding. Lab Chip 5:1393–1398
18. Zare RN, Luo YQ, Yu F (2008) Microfluidic device for immunoassays based on surface plasmon resonance imaging. Lab Chip 8:694–700
19. Wei H, Chueh BH, Wu H, Hall EW, Li CW, Schirhagl R, Lin JM, Zare RN (2011) Particle sorting using a porous membrane in a microfluidic device. Lab Chip 11:238–245

Chapter 5
Cell Co-culture and Signaling Analysis Based on Microfluidic Devices Coupling with ESI-Q-TOF MS

5.1 Introduction

The cell analysis is carried out with the intention to explore the physiological activities of living organisms, therefore obtain useful information on of disease prevention and treatment. The living organism is an organic entity, and each organ exists interdependently. Every single external stimulation would cause a chain reaction of the related organs, which induce the whole body's quick response. The signal transmission between organs essentially is the transmission of certain substance, which was secreted by the cells belonging to the stimulation-caused organ. This compound searches for the receptor on the target tissue, and further transfers the signal. Through regulating the content and type of the secretions acting as the signal factors, the information is transferred to the receptor cells, and response is made by the living organism. Therefore, cell signaling is a part of a complex system of communication that governs basic cellular activities and coordinates cell actions, which requires for further studies. The lack of perceiving and correctly responding to their microenvironment is one possible reason for the loss of functional capabilities of development, tissue repair, and immunity as well as normal tissue homeostasis, which causes diseases such as cancer, autoimmunity, and diabetes.

In conventional biological methods, only individual tissues or types of cells were investigated, but no attention was paid on the native cell-cell signaling process. The signal transmission between the cells was cut off since the tissues were separated from the living organism, which blocked the way to explore the chain reaction of related cells under stimulation. In order to solve this problem, multiple types of cells co-culture methods were developed recently to better mimic the organization and complexity of the *in-vivo* microenvironment. Due to the micro-scale structures and precise control of the chemical environment, microfluidic devices were verified for the increasing capability on cell analysis. Several works on the multiple types of cells co-culture on microfluidic devices were reported to mimic the *in-vivo* environment. Most studies focused on fabricating highly complex, well-organized,

H. Wei, *Studying Cell Metabolism and Cell Interactions Using Microfluidic Devices Coupled with Mass Spectrometry*, Springer Theses, DOI 10.1007/978-3-642-32359-1_5,
© Springer-Verlag Berlin Heidelberg 2013

two-dimensional (2D) or three-dimensional (3D) microscale structures. Surface pattern [1], assembled substrates [2], and both sides of a modified porous membrane [3] have been reported to achieve co-cultures of different types of cells. ECM [4] and collagen [5] was widely selected as the scaffold materials to build a 3D environment which could hold the cells in network structures, as well as some attempts on other substitutes. However, up to now efforts on cells co-culture studies have continued to investigate only one factor at a time, with limited detection approaches, such as fluorescence imaging. The observation mainly focused on the cells migration and quantity change caused by one factor, which was not sufficient for thorough determination of multiple cells signaling pathways. The approach to determine the composition and content of the cell signaling factor is further required by research on cell communication.

In this chapter, based on the developed cell analysis platform, we present an approach to structure a micro-environment for the co-culturing of neuron cells and pituitary cells to simulate the cell signaling inside a mammalian organism. Growth hormone (GH) released by the anterior pituitary gland is a protein-based peptide hormone, which stimulates growth, cell reproduction and regeneration in humans and other mammalian animals. The most common disease of GH excess is a pituitary tumor composed of somatotroph cells of the anterior pituitary. The abnormal release of GH mostly causes dwarfism, gigantism, and acromegaly. Its synthesis and release is under tonic inhibitory control by a neurotransmitter, which was secreted from the neuron cells. The model cell signaling pathway of neurotransmitter regulating the GH release from pituitary cells was investigated, because of its essential function in the life activities.

In the experiment, GH3 (rat pituitary tumor cells) and PC12 cells were co-cultured in two connected microfluidic channels, which allowed the signal factors diffusion but forbidden different types of cells mixture. A micro-SPE column was integrated in order to remove salts from the cells secretion prior to mass spectrometry detection. The micro valves were integrated in a three layer PDMS microfluidic device to avoid contamination between the cells co-culture zone and the pretreatment zone. The inhibition effect for rGH secretion from GH3 was investigated from the regulation function of dopamine released from PC12 cells. In this work, the cell analysis platform was further integrated to be applied on manipulating the extracellular environment of co-culturing cells, collecting the secretion products released by the cells, and determining the secretion products with a sensitive analyzer for medical screening tests or rapid diagnosis.

5.2 Experimental Section

5.2.1 *Reagents and Materials*

Negative photoresist (SU-8 2050 and SU-8 2007) and the developer were obtained from Microchem Corp. (Newton, MA, USA). Silicon wafers were purchased from

Xilika Crystal polishing Material Co., Ltd. (Tianjin, China). Positive photoresist AZ50XT and the developer AZ 400K were purchased from AZ Resist (Somerville, NJ, USA). Poly(dimethylsiloxane) (PDMS) and the curing agent were obtained from Dow Corning (Midland, MI, USA). Poly-L-lysine was purchased from Sigma-Aldrich Chemical Co. (St. Louis, MO, USA), which was applied to modify the glass slides surface to get a better cell adhesion. Methanol (HPLC grade) was obtained from Fisher Scientific (Springfield, NJ, USA). Live/dead viability/cytotoxicity assay kit (Invitrogen, CA, USA) was used to test the viability of the co-cultured cells. The cell membrane labeling solution DiI and DiO for the distinction of different cell types were obtained from Vybrant (Invitrogen, CA, USA). The packaging material for the pretreatment of the cell secretions was obtained from the SPE C18 (macropore) column (Agela, Tianjin, China). Tyrosine and norketamine, which were used for the stimulation of PC12 cells, were purchased from J&K chemical company (Beijing, China). The recombinant rat growth hormone (rGH) was obtained from Prospec-Tany (Rehovot, Isreal). Methotrexate (MTX), obtained from Fluka Chemicals (Castle Hill, NSW, USA) is an anti-cancer medicine.

The plasma cleaner PDC-32G (Harrick Plasma, Ithaca, NY, USA) was applied for oxygen plasma treatment. The syringe pump KDS100 (kdScientific, Holliston, MA, USA) was used to deliver eluting solutions in accurate rates. A fluorescence microscope (Leica DMI 4000 B, Wetzlar, Germany) equipped with a CCD camera was used to observe and to take images of the newly fabricated microfluidic devices. By using Leica Application Suite, LAS V2.7 the dimensions of each microstructure could be manually measured. For the valve alignment the stereomicroscope XSZ-G (COIC, Chongqing, China) with the extended light source Leica CLS 100X (Leica, Wetzlar, Germany) was used. A 500 µL syringe was obtained from Hamilton (Bonaduz AG, Switzerland). ESI-Q-TOF detection was carried out with a Bruker microTOF-Q mass spectrometer (Bruker Daltonics Inc., Billerica, MA, USA). All mass spectra were obtained in the positive mode. Mass spectra analysis was performed with DataAnalysis (Bruker Daltonics Inc.).

5.2.2 Design and Fabrication of Soft Lithography Mold

The integrated microfluidic device was built up of three layers: the PDMS valve control layer, the PDMS flow layer, and the glass substrate. The PDMS layers were obtained by replicate molding on silicon wafers. As shown in Fig. 5.1, a two-step photolithography technology was applied to structure microchannels with different heights in the PDMS flow layer. Briefly, on a silicon wafer cleaned by piranha solution, the first layer patterns were produced by spin coating the negative photoresist SU-8 2007 with a speed of 3,000 rpm to generate a 7 µm thick film. After exposure and development under UV light, the already patterned wafer was coated by SU-8 2050 using the same rotation speed to obtain a 60 µm thick film for a higher structure. In order to avoid an uneven surface, the wafer was loaded for 30 min before a second exposure.

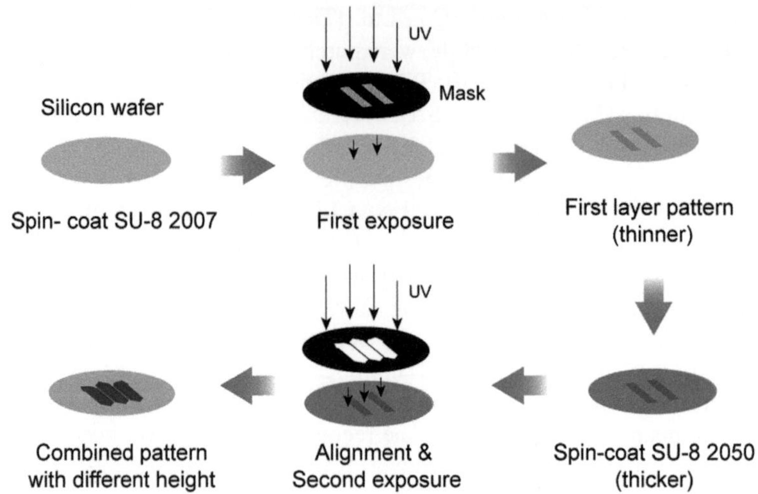

Fig. 5.1 The procedure of two-step photolithography technology which was used to structure microchannels with different heights in the PDMS flow layer

After the production of the second layer, the first layer pattern was still visible, since the photoresist is transparent. With the aligning markers designed on the mask, the first and second layer patterns were adjusted perfectly under the microscope.

The valves mold fabrication required a positive photoresist. Thus, the third layer pattern was fabricated after the photolithography patterning process. To produce the valve controlled flow channels, the positive photoresist AZ 50XT was spin-coated at a speed of 2,000 rpm onto the patterned wafer to obtain a 40 μm film. Alignment and exposure steps were carried out as described before. The developer AZ 400K was diluted by deionized water at a 1:4 ratio for a sufficient development of the photoresist. The prepared mold wafer was placed on a heating plate at 120°C for 4 min, because a further heating procedure was required for the AZ 50XT photoresist reflowing to construct a curved surface. It is known from literature that the round profile is critical for complete valve closure [6]. In contrast, the shape of the channels made of negative photoresists (SU-8 2007 and SU-8 2050) cannot be changed thermally, because SU-8 forms a rigid structure after photopolymerization by ultraviolet light. In the case of producing structures with multiple patterns on the same wafer, the order of photoresist coating should be carefully designed by considering the compatibility of developers and photoresists. As an example, in our experiments AZ 50XT can be removed by the SU-8 developer, which could cause severe problems to the three dimensional structure.

After the third and final exposure and development, a mold carrying a relief of the desired microstructure was fabricated. The mask design and alignment marks of the combined three layer patterns are shown in Fig. 5.2.

The mold for fabricating the valve control layer was prepared by coating SU-8 2050 at a speed of 3,000 rpm on a silicon wafer, to form a film with a height

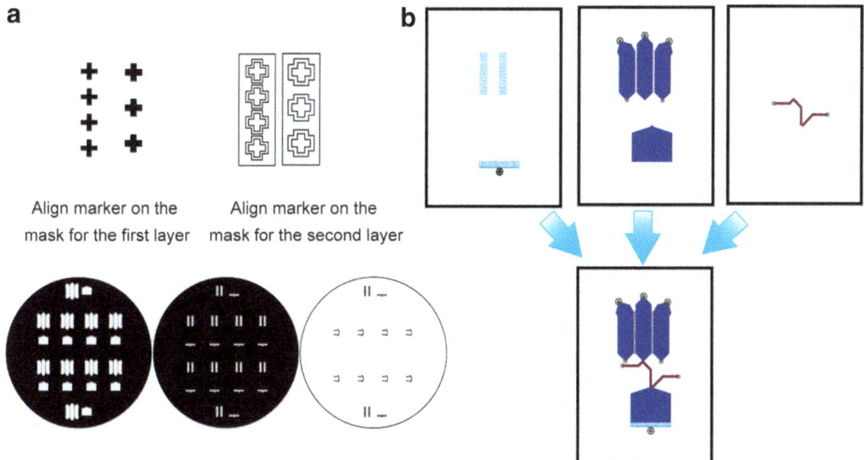

Fig. 5.2 The mask design and alignment markers of the three layers combined pattern. (**a**) Alignment marks and the mask designs. (**b**) Scheme of the fabrication of the three layer patterns, with the final combination of all patterns

of 55 μm. Prior to their use in soft lithography, all molds were incubated with 1H,1H,2H,2H-perfluorooctyl trichlorosilane vapor in a vacuum desiccator for 2 h in order to prevent adhesion of the cured PDMS and the mold.

5.2.3 Design and Fabrication of Microfluidic Devices

The microfluidic device included three layers: the valve control layer, the microchannel layer, and the glass slide. In order to fabricate a 4 mm thick piece of PDMS for the valve control layer, a 5:1 premixed PDMS prepolymer (elastomer: curing agent) was prepared, degassed in a vacuum chamber, and then poured on the mold and finally cured in a 75°C oven for 30 min. Meanwhile, a 20:1 mixture (elastomer: curing agent) was spin-coated at a speed of 2,000 rpm on to the flow channel mold, and cured at 75°C for 15 min after a loading time for 20 min. The valve control layer PDMS was cut with a surgical scalpel and then gently peeled off the mold. Holes were generated by a flat-tip syringe needle to serve as an inlet for pressurized air to close the flow channels. The alignment of the valve layer with the flow layer was achieved under a stereomicroscope with an extended light source. To form one integrated PDMS microchip, the aligned layers were cured in the 75°C oven for 60 min. In the next step the cured PDMS layers were peeled off the thin layer mold and a sample access hole was punched through the layers in order to form the cells and packing materials inlets and outlets. Finally, the prepared PDMS layers were covered with a glass slide irreversibly after an oxygen plasma treatment for 90 s. A modification of the cell co-culture microchannels with poly-L-lysine allowed for a better cell adherence.

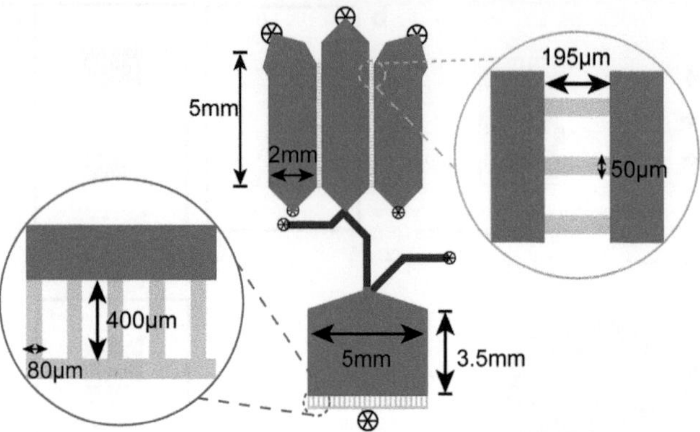

Fig. 5.3 Schematic illustration for the measurements of the three main areas in the microfluidic devices

As shown in Fig. 5.3, the microfluidic device was composed of three main zones: the cell co-culture zone, the secretion pretreatment zone, and the flow control zone. The cell co-culture zone contained three main cell culture channels, and two groups of minor connecting channels. The main cell culture channels were produced with a length of 5 mm, a width of 2 mm, and a height of 60 µm. These dimensions allowed for a smooth loading and sedimentation of the cells. The minor connecting channels were fabricated with a length of 195 µm, a width of 50 µm, and a height of 7 µm, in order to prevent a hybridization of loaded cells between different inlets. Furthermore, the connection was large enough to ensure that the chemicals diffusion between the neighbor channels was possible. The channel in the pretreatment zone was designed to be 3.5 mm long, 5 mm wide, and 60 µm high. The end of the pretreatment channel was connected with the outlet consisting of a group of small channels, which had a length of 400 µm, a width of 80 µm, and a height of 7 µm. The small channels were designed to act as a dam for trapping the pretreatment packing materials, which had an average diameter larger than 7 µm.

As shown in Fig. 5.4c, the flow control function was achieved by two PDMS layers. The red channels were set in the bottom layer, with a width of 300 µm and a height of 40 µm, the yellow channels were set in the top layer, with a width of 420 µm and a height of 55 µm. The schematic illustration of the microfluidic device is shown in Fig. 5.4.

5.2.4   General Cell Culture and Staining

GH3 cells (Runcheng Biotechnology Co., Ltd, Shanghai, China) were incubated in pH 7.4 growth medium consisting of DMEM supplemented with 10% heat-inactivated fetal calf serum, 100 U/mL penicillin G, and 100 mg/mL streptomycin.

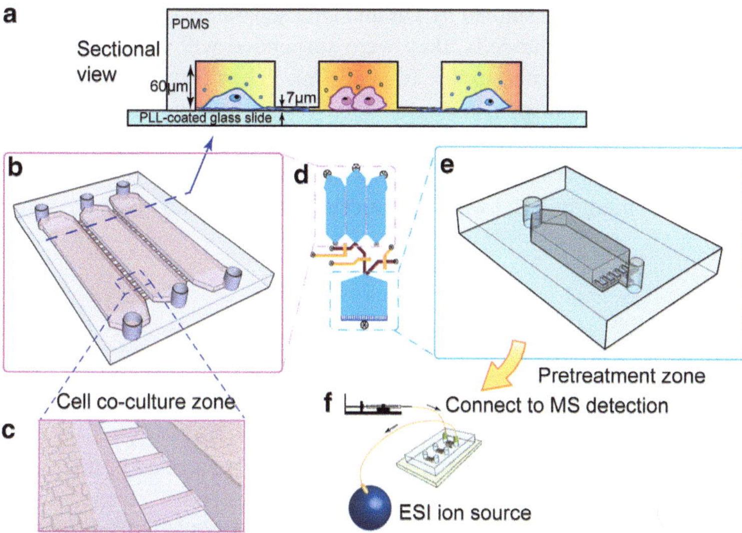

Fig. 5.4 Schematic illustration of the microfluidic device. (**a**) The sectional view of the cell coculture zone cut from the *dashed line* in (**b**). Different types of cells could be co-cultured, with the signal factor communicating by diffusion. (**b**) Scheme of the cell co-culture zone. (**c**) A magnified illustration of the connected small channels between the adjacent cell culture channels. (**d**) The overall schematic representation of the microfluidic chip with cell co-culture zone, control valves, and pretreatment zone. (**e**) Scheme of the pretreatment zone. (**f**) Scheme for the coupling of the microfluidic device to the mass spectrometer for protein detection (Reprinted with permission from Ref. [8]. Copyright 2011 American Chemical Society)

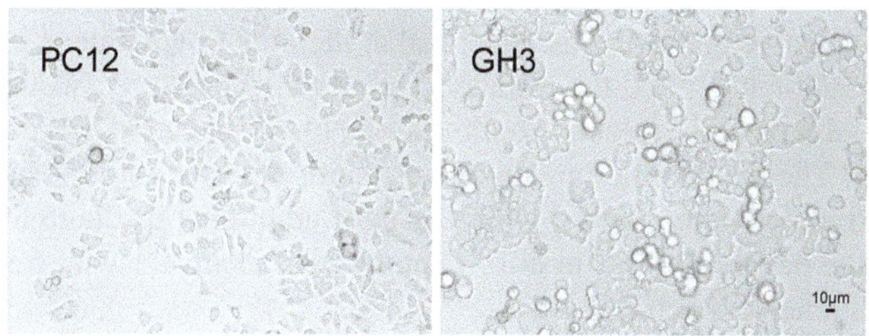

Fig. 5.5 The microscope images of PC12 and GH3 cells during normal culture conditions

The PC12 cells (Beijing Tumor Hospital, Beijing, China) required 5% horse serum. Both cell types were cultured in a 37°C humidified incubator with 5% CO_2. Subcultures were performed when the cells reached confluence. Cells were trypsinized at a ratio of 1:3 after confluence by using a 0.25% trypsin solution. The microscope images of PC12 and GH3 cells are shown in Fig. 5.5, which were taken under normal culture conditions. PC 12 cells adhered to the substrate tightly, while their shape was irregular. GH3 cells adhered in a nearly round shape, and tended to assemble as multiple layers.

Subcultured cells were stained by DiI and DiO separately to distinguish different cell types. For staining the cells, they were suspended with DMEM solution and the cell density was adjusted to 1×10^6 cells/mL. Then 5 μL of the cell-labeling solution per mL of cell suspension were added. Afterwards the cells were incubated at 37°C for 20 min. Finally the labeled suspensions were centrifuged at 1,500 rpm for 5 min, and the supernatant was removed and the cells were gently resuspended in warm (37°C) medium. The washing procedure was repeated twice to remove all residual dye.

5.2.5 Co-culture of PC12 and GH3 Cells on Microfluidic Device

After being cultured and stained, the adherent cells were seeded in microchannels, which were coated by poly-L-lysine. Glass surfaces were cleaned in piranha solution, rinsed with deionized water, then put into a 100% ethanol solution for 30 min, and dried with nitrogen. The PDMS was finally irreversibly sealed to a commercially available microscopy glass slide. The coating with poly-L-lysine was done by applying a 0.1% poly-L-lysine solution to the microchannels. The poly-L-lysine solution was incubated with the glass surface for 4 h, rinsed with PBS buffer and water for three times each, and finally dried at room temperature. Prior to the cell seeding step, the entire microfluidic device was sterilized with ultraviolet radiation in a super clean bench for at least 30 min.

The cells were trypsinized, and resuspended at a density of about 10^6 cells/mL before seeding. 2 μL cell suspension were added into the cell culture channel inlet, and a negative pressure was generated at the outlet by a pipette until the cell suspension completely filled the channel. The negative pressure was used to avoid a cell accumulation at the inlet and outlet caused by the injection process in the cell seeding step. Then cell culture medium was gently injected into the channels and the inlet and outlet were covered with a drop of additional media to avoid drying up. At last the device was put in a 37°C humidified incubator with 5% CO_2. The medium was renewed each 8 h.

In our experiments, the designed three main cell culture microchannels (Fig. 5.4a) were used to culture PC12 cells in channel 1 and channel 3, with GH3 cells cultured in channel 2. This design allowed for better neurotransmitter diffusion into the GH3 cell culture channel. Since the suspended cells had a spherical shape, with diameters in the range of 16–22 μm, which is larger than the diameter of the minor microchannels connecting the neighbor cell culture channels, different types of cells could not hybridize in the following steps. All further experiments were carried out within 24–48 h after cell seeding. In the drug exposure experiments the cell culture medium was replaced against PBS buffer.

5.2.6 Drug Exposure

PC12 cells were incubated with L-tyrosine (1 mM) dissolved in PBS buffer for
3 h before the secretion products of the GH3 cells were detected. As a control
experiment L-tyrosine solution was injected into channel 1 and 3 without culturing
PC12 cells. Norketamine was selected to carry out a comparative trial. Norketamine
is a nerve narcotic drug with an efficient effect, but it is also known to have a
short half-life. Norketamine hydrochloride was dissolved in PBS buffer to a final
concentration of 0.1 mM. In the comparative trial the solution within the cell culture
channels was changed against freshly prepared norketamine solution each half an
hour for six times.

5.2.7 Pretreatment Procedure

An integrated C18 silica packing material micro column in the microfluidic device
was used to pretreat growth hormone collected from the GH3 cells supernatant.
After being isolated from the SPE C18 (macropore) column, the packing material
was flushed with 50% methanol solution and suspended by vortexing. The suspension was injected into the microchannels with the valve closed, that controls
the microchannel connecting the cell culture part and pretreatment part. The
capped macropore C18 silicon are reversed phase extraction materials, with a high
bonding density and very efficient recovery under certain conditions. Furthermore,
macropores are more suitable for the desalination and concentration of protein,
DNA and other macromolecules. However, the particle distribution of the used C18
silica was quite wide, so a very low dam was fabricated to trap the packing materials,
and a group of parallel microchannels was designed at the end of the micro SPE
column to reduce the flow rate and provide a sufficient contact between the analyte
and C18 particles.

The micro SPE column was activated by methanol (for packing the materials)
then followed by a washing step with 100 μL deionized water. The sample from the
cell culture channel was added (less than 1 μL), followed by a washing step with
100 μL 20% (v/v) methanol solution, and then eluted by 80% methanol containing
0.1% formic acid for mass spectrometry detection, while the consumption of eluting
solution was less than 10 μL.

5.2.8 Mass Spectrometry Settings

A quadrupole ion trap mass spectrometer with a modified micro-ESI source was
used for the protein detection. The heated inlet capillary was set to 200°C. A coaxial
nebulizer nitrogen gas flow (0.4 bar) around the ESI emitter was used to support

the generation of ions. The electrospray voltage was 4.5 kV in the positive ion mode. All mass spectra were externally calibrated by Tunemix (Agilent, USA) in the positive ion mode within a m/z range of 1,500–3,000.

5.3   Results and Discussion

The overall goal of this work was to design a microfluidic device, which can simulate the signal communication between different tissues, collect and purify the released secretions, and analyze the secretion products qualitatively and semi-quantitatively. Here we describe the communication of different cells as this is the basic of the communication between different kinds of tissue. The chosen model is the neuroendocrine regulation for the pituitary release of growth hormone. As shown in Fig. 5.6, the neurotransmitters secreted from PC12 cells are ingested by GH3 cells after passing the connected minor channels. Thereby, the growth hormone release behavior of GH3 cells was regulated as a function of neurotransmitters.

In the present work dopamine was used as a neurotransmitter substance. Dopamine is biosynthesized in the PC12 cells first by the hydroxylation of the amino acid L-tyrosine to L-DOPA via the enzyme tyrosine 3-monooxygenase, and then by the decarboxylation of L-DOPA by aromatic L-amino acid decarboxylase.

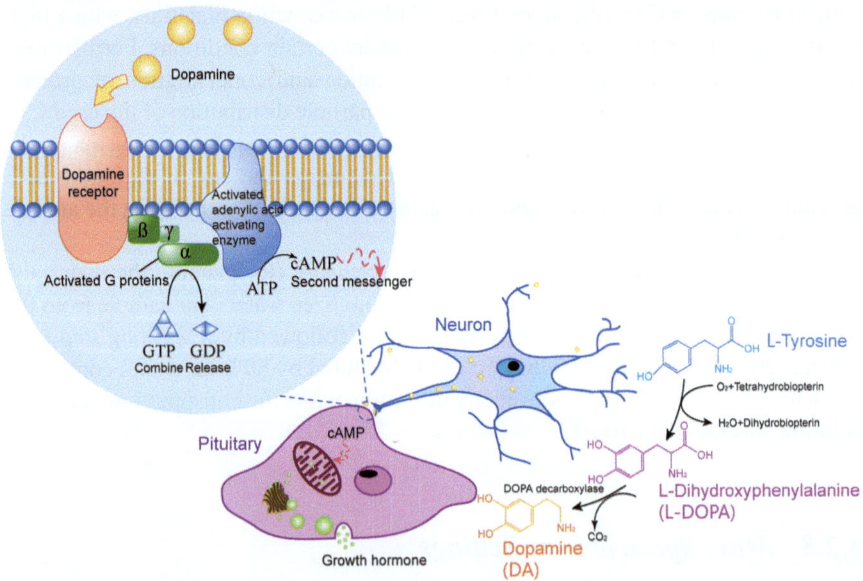

Fig. 5.6 Scheme shows the synthesis of dopamine as one of the neurotransmitters in the neuroncell. The growth hormone release behavior of pituitary cells was regulated as a function of neurotransmitters (Reprinted with permission from Ref. [8]. Copyright 2011 American Chemical Society)

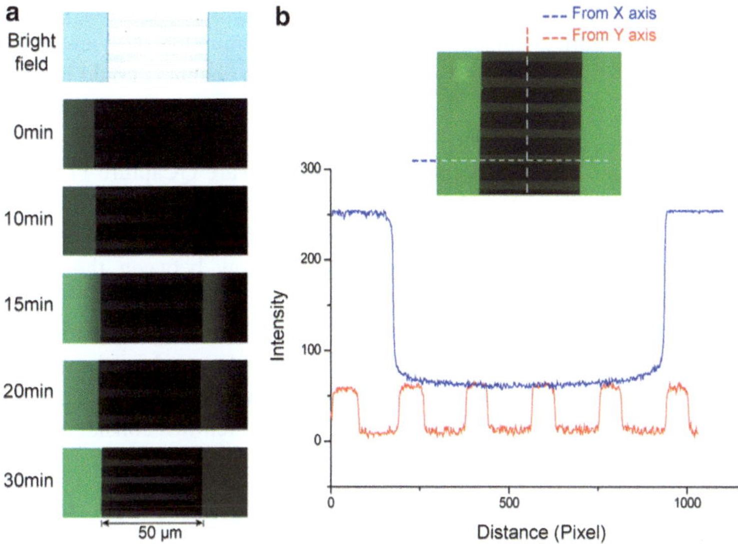

Fig. 5.7 Evaluation of the mass diffusion rate in the co-culture microfluidic channels. (**a**) Microscopic fluorescence images obtained from 0 to 30 min after fluorescein sodium salt solution was injected into the *left channel*. (**b**) Fluorescence intensity evaluation along the two *dashed lines* after the diffusion reached the equilibrium (Reprinted with permission from Ref. [8]. Copyright 2011 American Chemical Society)

After being released to the extracellular environment, dopamine binds to a specific receptor on the GH3 cells surface. The G protein-coupled receptor is then activated as a transmembrane receptor. The transformation between guanosine triphosphate (GTP) and guanosine diphosphate (GDP) switches the G proteins function and causes the activation of the adenylate cyclase, which can catalyze ATP to cAMP as a second messenger to direct the secretion downstream in the cell [7].

5.3.1   Evaluation of the Mass Diffusion Rate in the Co-culture Microfluidic Channels

Since the different types of cells were cultured in adjacent channels, and the mass delivery was achieved only by diffusion, it's important to evaluate the mass diffusion rate between the cell cultured microchannels, which were connected by the minor lower bridge channels. A 1 mM fluorescein sodium salt solution, with the molecular weight of 376.27 g/mol which is in the same order of magnitude as the dopamine's molecular weight of 153.18 g/mol, was adopted to investigate the diffusion rate. As shown in Fig. 5.7, food dye colored solutions were injected into the three parallel channels 1, 2, and 3 at the same time, with 1 mM fluorescein sodium salt in the solution filled into channel 1 and 3. The injection was immediately stopped after the channels were completely filled.

The microscopic image is shown in Fig. 5.7a. The connection part of two neighbored cell culture channels was presented and limited to the viewing field of the microscope. A fluorescence balance between two channels could be observed 30 min after the injection of the fluorescein solution. The fluorescence distribution was analyzed on the picture taken 40 min after the solution was injected. Afterwards the fluorescence intensity was evaluated by the software QCapture Pro 5.1 along the dashed lines shown in Fig. 5.7b, from both X axis and Y axis. The red curve indicates that the six parallel connecting channels showed the same fluorescence intensities, while the background was extremely low between the connecting channels. The blue curve displays the fluorescence intensity of the cell culture channels was similar, while the fluorescence intensity in the connecting channel area was significantly lower, since these are much lower than the cell culture channels. The results demonstrated a completely mass diffusion balance between the neighbored cell culture channels within 40 min, which was considered to be the incubation time in the following drug application experiments.

5.3.2 Mass Diffusion Effect on the Cell Culture

In the life entity, the extracellular environment is not always steady. The mass concentration is changing to transfer a specific signal from the upstream of the regulation chain. To mimic this extracellular environment, the connected cell culture microchannels were designed in a special way. First, each type of cells was cultured separately in one of the connected microchannels to evaluate the mass diffusion effect on the cell culture.

Single type cells were cultured in the left channel, while the right channel was filled with serum rich culture medium. As shown in Fig. 5.8, the cells position changed after being cultured for 20 h and 36 h. The red dash line marked the displacement. In Fig. 5.8a and b, the cell 1 moved forward about 8 μm, while the cell 2 moved forward for 4 μm. In Fig. 5.8c and d, the presynaptic ending moved 25.6 μm to the microchannel with no cells cultured.

Although the cells seeded into the neighbored microchannels could not hybrid since the cells diameter was larger than the height of the connecting channels when they were suspended. Interestingly, the cells shape could change to a more flat shape after adhering to the ground. PC12 cells preferred to extend more open, and showed a much thinner shape than in the suspended situation, which can explain a faster migration in the connected channels. After being seeded in the microchannels, consumption of the nutrition media and air in the culture medium decreased with the cells propagation. Thus, the cells tended to migrate to an area with more sufficient nutrition, air, and space. Although the nutrition and the air were diffusing into the cell cultured channel all the time, the cells consumption was constantly on a high level, which caused a migration of cells to the empty microchannel.

In order to further demonstrate the mass diffusion effect on the cell culture in the newly designed device, 5 μM methotrexate (MTX) was added into the

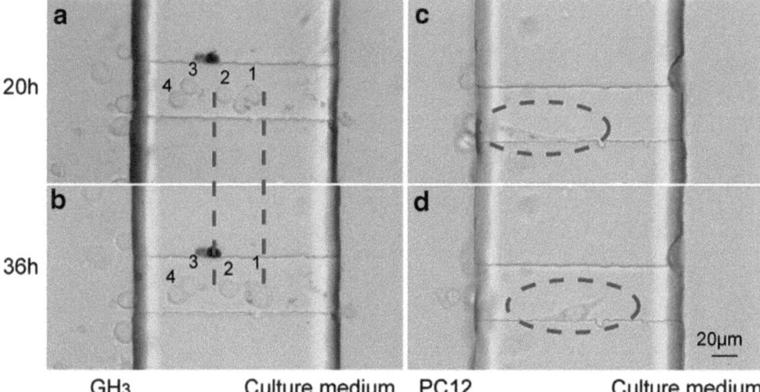

Fig. 5.8 Microscopic images show the cells migration due to the mass diffusion between the connected microchannels. (**a**) and (**b**) show GH3 cells, while (**c**) and (**d**) show PC12 cells (Reprinted with permission from Ref. [8]. Copyright 2011 American Chemical Society)

culture medium to fill the empty micro channel. As an anti-cancer medicine, MTX competitively inhibits the dihydrofolate reductase (DHFR), which is an enzyme participating in the tetrahydrofolate synthesis, to inhibit the growth and reproduction of the tumor cells. Figure 5.9 shows the cells growing situation after changing the culture medium to the MTX solution. To keep the MTX concentration constantly on the same level, the MTX solution was refreshed each hour. After applying MTX for 2 h, the edge of the GH3 cells cluster started to draw back (as the dashed lines show). After 4 h, the cell density started to decrease and after 8 h, the cell density reduced clearly, and numerous dead cells were found in the inlet and outlet. Figure 5.9d shows when a group connecting channels with a lower inter-channel distance was constructed, the cell density declined even more within 8 h. The cells migration and declining demonstrated the mass diffusion effect on the cell culture in the extracellular environment, and proved the further effect on the cells co-culture in the designed device, since the signaling factors could affect the receptor cells due to mass diffusion.

5.3.3   *PC12 and GH3 Cell Induction Under Co-culture Conditions*

The PC12 and GH3 cells were seeded into the three microchannels separately with a density of 10^6 cells/mL as described before, in order to create a faster signal factor diffusion.

The growing and reproduction situation of co-cultured PC12 and GH3 cells were as active as cultured in the sterilized dishes (Fig. 5.10). The pH 7.4 growth medium

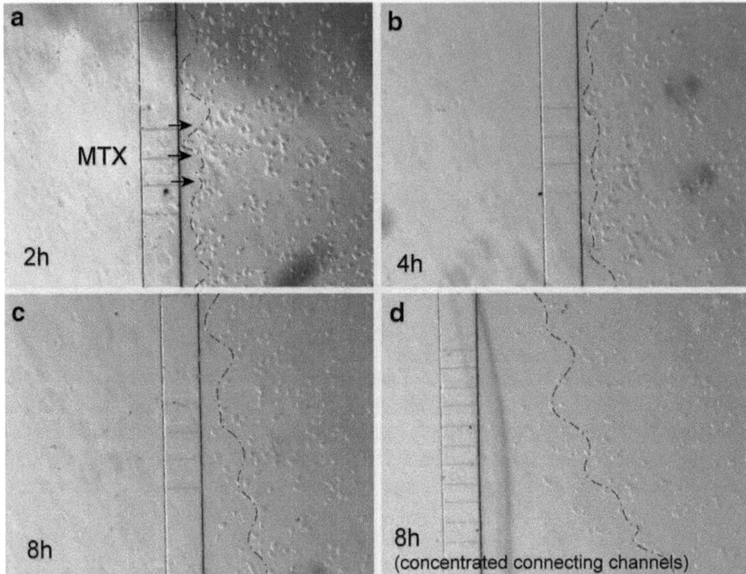

Fig. 5.9 Cell growing conditions when 5 μM methotrexate was constantly added into the *left channel* for (**a**) 2 h, (**b**) 4 h, and (**c**) 8 h. The *dashed lines* show the rough edge of the GH3 cell clusters. (**d**) Cell growing condition when the connecting channels with a lower inter-channel distance was used (Reprinted with permission from Ref. [8]. Copyright 2011 American Chemical Society)

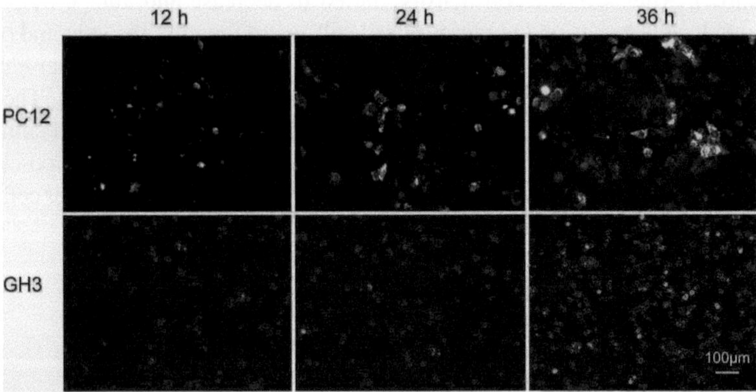

Fig. 5.10 Growing and reproduction situation of co-cultured PC12 and GH3 cells in the newly constructed microfluidic device. The growing speed within the microchannels was as fast as that of the cells cultured in the sterilized dishes. The cell membrane dyes DiI and DiO were applied to distinguish different types of cells. The staining ability of DiO (*green fluorescence*) is weaker than DiI (*red fluorescence*). Additionally, the PC12 cells extended more under the adherent growing condition, so the intensity of the *green fluorescence* was not as strong as the *red fluorescence*

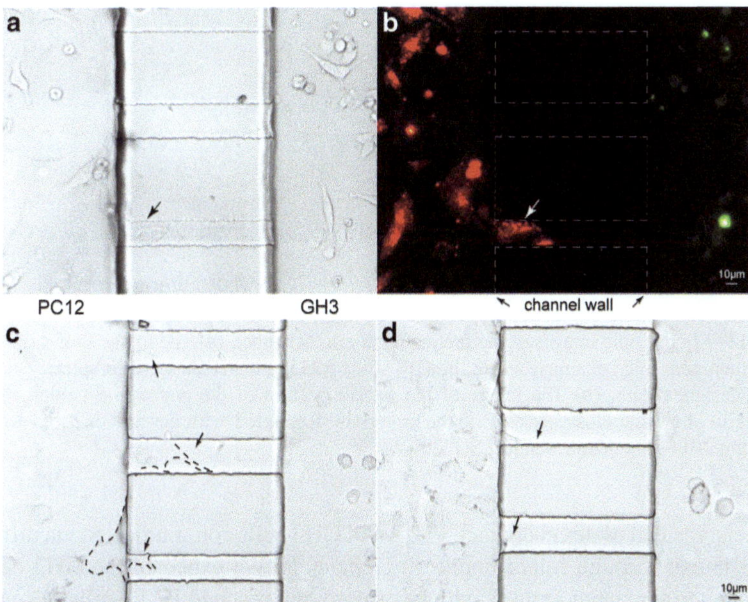

Fig. 5.11 PC12 and GH3 cells co-cultured in the newly designed microfluidic device, and PC12 cells migration was observed. (**a**) and (**b**) The *bright field* and *fluorescent image* show the migration of PC12 forwards to GH3 cells. (**c**) and (**d**) Images of co-cultured PC12 and GH3 cells show the migration phenomenon (Reprinted with permission from Ref. [8]. Copyright 2011 American Chemical Society)

consisting of DMEM supplemented with 10% heat-inactivated fetal calf serum was used as the common culture medium for both types of cells. The cell staining dyes DiI and DiO were applied on different types of cells, to distinguish PC12 and GH3 cells, since the membrane fluorescence mark from the mother cell was also present in the daughter cells after the karyokinesis.

As discussed in Sect. 5.3.2, a migration from the cell culture microchannels to the empty channels was observed when a single type of cells was cultured. After both sides of the connected channels were filled with cells, the consumption of the nutrition and air were equal to each other, and the migration was supposed to stop. However, the migration and synapse extending was observed when the PC12 and GH3 cells were co-cultured in the connected microchannels. As shown in Fig. 5.11, PC12 cells were cultured in the left channels with red fluorescence, while GH3 cells were cultured in the right channels with green fluorescence. The PC12 cell migrated forwards to the right side through the connected channel, tended to be closer to the GH3 cultured zone. Figure 5.11c and d show another two cases of the migration of the PC12 cells to the GH3 cell culture zone. As a control experiment, HepG2 (human hepatocellular carcinoma) cells were cultured instead of PC12 cells. The migration seen with the PC12 cells was not observed with the HepG2 cells. To sum up, in this study we patterned PC12 and GH3 cells in two separated cell culture

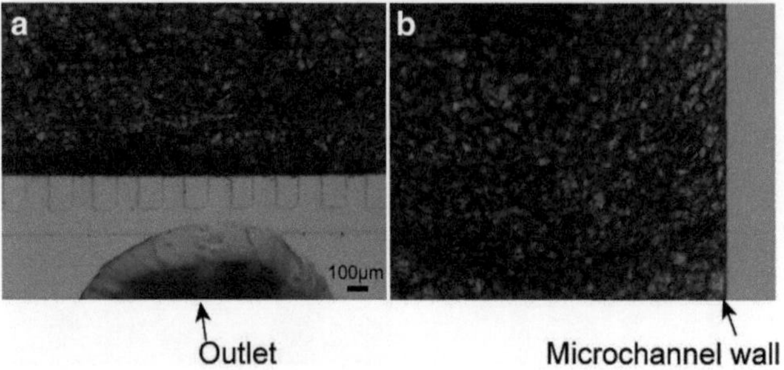

Fig. 5.12 Microscopic images of the pretreatment microchannel. (**a**) The image shows the end of the pretreatment microchannel, where the C18 silica packing materials were completely captured by the designed dam. (**b**) The image of the middle section of the pretreatment microchannel, in order to show the close packing of the materials (Reprinted with permission from Ref. [8]. Copyright 2011 American Chemical Society)

channels. Instead of direct contact, PC12 and GH3 cells communicated via diffusion of substances through minor connecting channels. We expected the GH3 cells to generate certain signal factors, which diffused and reached PC12 cells to promote the nerve synapse's extension forwards to the receptors on the GH3 cells surface.

5.3.4   Evaluation of the Micro-SPE Column

Measuring the binding capacity of the micro-SPE column is important for the pretreatment evaluation as well as for the determination of real samples. In order to investigate the capacity of the prepared micro-SPE column, rGH solutions with a series of concentrations were prepared. The protein rGH has a molecular weight of 21,980 Da.

As shown in Fig. 5.12, the pretreatment column was filled with a C18 packing materials, which had a wide particle size distribution. Although the size distribution is wide, the designed low dam completely trapped all the material, which also induced a high column back pressure. Thus slow flow rates of 2 μL/min were used in all experiments.

The micro-SPE column was prepared while the valve was closed, which was located between the cell culture zone and pretreatment zone, in order to prevent contamination. First the C18 packing materials were wetted with methanol and suspended by vortexing. Afterwards the suspension was injected into the prepared microchannels to pack the micro-SPE column until a defined position was reached. Finally the miniature SPE column was pre-conditioned by flushing it with 100 μL deionized water. For the measurement of real samples, the sample was induced from cell culture channel 2 into the micro-SPE column, by opening the valve in the middle and closing the other two valves. During this process, the inlets and outlets

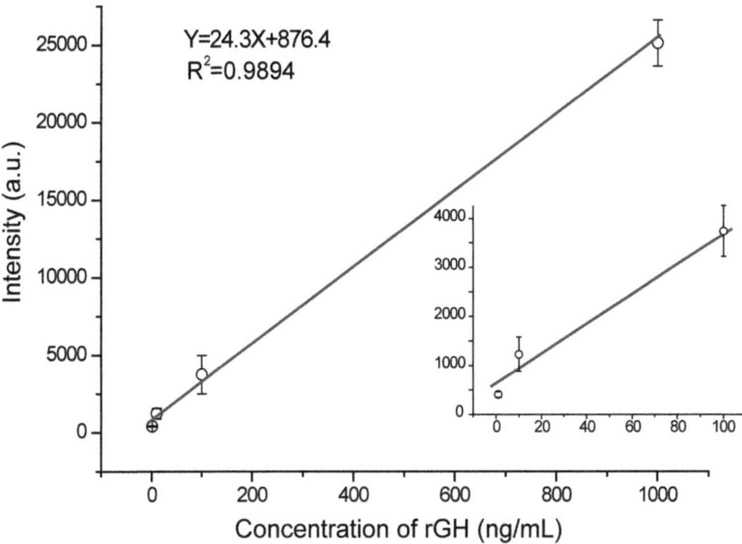

Fig. 5.13 The *calibration line* shows the relationship between the intensity of the area underneath the peaks of the mass spectra and the concentration of rGH. The sum of the intensity of *m/z* 1,691, 1,833, 1,999, 2,199, and 2,443 was calculated as the y-coordinate. Every plot was the average of five repeated measurements (standard deviations are indicated as error bars) (Reprinted with permission from Ref. [8]. Copyright 2011 American Chemical Society)

of channel 1 and 3 were sealed with stainless steel columns. The micro-SPE column was washed using 100 μL 20% ethanol, and then eluted by 80% methanol solution containing 0.1% formic acid, for direct injection into the mass spectrometer.

The rGH solutions were prepared in PBS buffer with final concentrations of 1, 10, 100, and 1,000 ng/mL. 3 μL sample solution were injected each time. In order to ensure the reproduction, the sum of the intensity of *m/z* 1,691, 1,833, 1,999, 2,199, and 2,443 were considered to generate a calibration line (shown as Fig. 5.13). The calibration line obtained on plotting the peak areas was linear in the range from 1 to 1,000 ng/mL and the linear equation was $Y = 24.3X + 876.4$ with an R^2 of 0.9894. The experiments were carried out five times to evaluate the precision of the assay. The evaluation results revealed that the integrated micro-SPE column has the capacity of growth hormone solution with the concentration of 1,000 ng/mL from the cell culture microchannel, which is much higher than the concentration produced by GH3 cells. In the following experiments, the capacity of the home-made micro-SPE column is 3 ng. In order to further evaluate the recovery for the detection of rGH, 50 ng/mL rGH solution was added into blank buffer. The recovery in five times repeated experiments was in the range of 93–133%. Based on the determination of blank samples, the sensitivity of the MS detection for rGH was 0.36 ng/mL.

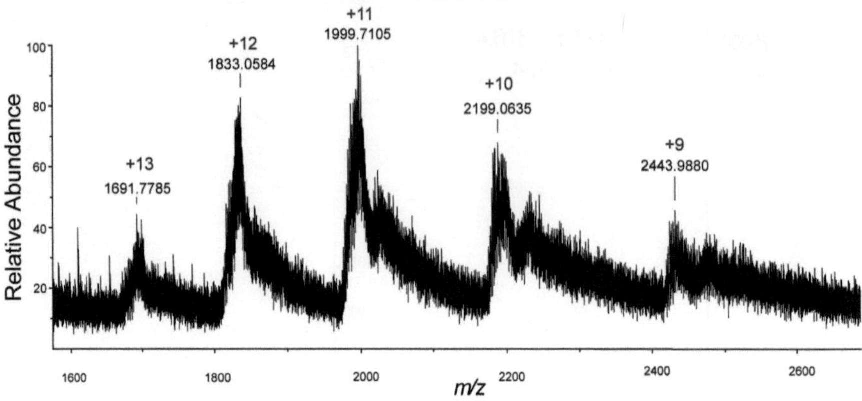

Fig. 5.14 Mass spectra of the rGH secreted by GH3 cells (Reprinted with permission from Ref. [8]. Copyright 2011 American Chemical Society)

5.3.5 Inhibition of rGH Secretion from GH3 Cells by Dopamine Released from PC12 Cells

The matrix of the GH3 cell secretion is composed of proteins, amino acid, and certain salts. In this environment the growth hormone production is remaining at a high level. Since the matrix of the normal culture medium is very complex, we replaced the culture medium with PBS in order to reduce matrix interferences. The rGH secreted from GH3 cells was analyzed by ESI-Q-TOF-MS after the pretreatment with a micro-SPE column. The mass spectrum obtained is shown in Fig. 5.14. According to the semi-quantitative standard curve, under culturing in PBS buffer, the rGH secreted by GH3 cells without co-culturing with PC12 cells within 3 h was collected and calculated to be as high as 69.9 ± 12.4 ng/mL.

In first control experiments, GH3 cells were cultured in channel 2 with and without HepG2 cells seeded in channel 1 and 3. In this section, all samples were collected for 3 h after changing to fresh PBS buffer. The left bar in Fig. 5.15 shows the rGH collected when only GH3 cells were cultured in channel 2, while the right bar shows the rGH obtained when HepG2 cells were co-cultured in channel 1 and 3. The comparison of the rGH secretion indicated, that HepG2 cells were not contributing to the rGH secretion regulation in GH3 cells.

The influence of the secretion of rGH by GH3 cells during the co-culturing with PC12 cells was investigated and discussed, as shown in Fig. 5.15. PC12 and GH3 cells co-culture experiments were carried out, whereas first a co-culturing for 24 h and for 48 h was made, followed by an exchange of the PBS buffer and another incubation for 3 h. The rGH, which was produced within 3 h was determined by ESI-Q-TOF-MS. The secretion of rGH was clearly decreasing after PC12 cells were co-cultured with GH3 cells, compared to the first experiment with HepG2 cells. The

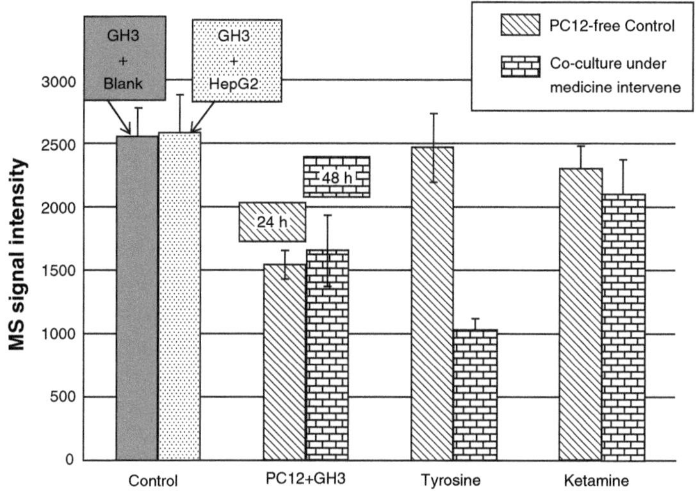

Fig. 5.15 The rGH secretion by GH3 cells during the co-culturing with PC12 cells (Reprinted with permission from Ref. [8]. Copyright 2011 American Chemical Society)

inhibition by PC12 cells didn't display a significant change from co-culturing for 24 h or for 48 h, which indicated the effect on GH3 from PC12 was on a constant level by signaling factor diffusion.

In order to prove that the neurotransmitter dopamine is the main reason of the rGH inhibition, another two control experiments were carried out. First, 1 mM tyrosine was applied on PC12 cells to promote the dopamine release, as the raw material for the dopamine synthesis. After 3 h application, the rGH secretion from GH3 cells was significantly declining. Another control experiment was carried out to avoid the effect on rGH secretion from tyrosine. 1 mM tyrosine was applied without PC12 cells co-cultured in channel 1 and 3. The results demonstrated that tyrosine had little influence on rGH secretion from GH3 cells, while dopamine was probably the reason of the decrease of rGH secretion.

To confirm the effect of dopamine during this process, 0.1 mM norketamine was applied on PC12 cells as a sympathetic block, which decreased the release of neurotransmitters. The contrast experiment was carried out by adding the same concentration of norkatamine into the same microchannel without PC12 cells being cultured. This microchannel was connected with the GH3 cell culture channel by minor channels. Compared to the results obtained by incubation with 1 mM tyrosine, the rGH secretion raised, but still didn't reach the concentration of rGH when GH3 cells were incubated without PC12 cells.

The above results confirmed the inhibitory effect on rGH secretion from GH3 cells by PC12 cells, and demonstrated the feasibility of the coupled platform between the microfluidic device and mass spectrometry for cell signaling studies.

5.4 Conclusion

In this chapter, we developed a microfluidic device which was coupled with mass spectrometry for various types of cells co-culture and signal factor analysis. A micro-SPE column was integrated into the chip in order to remove salts from samples obtained in biological environments, which is necessary for ESI-MS detection. This platform was proven to be a versatile and powerful tool to study cell signaling for various biological applications. It provides a well-controlled cell culture environment which can be adjusted more precisely by fabricating special structures. This combined system allows evaluating multiple biochemical factors, which are essential in mimicking physiological conditions, while cells constantly receive signals from soluble environments. Furthermore, the character and content of the released factor in the regulatory pathway was determined by the high sensitive MS. This cell co-culture platform would be very useful in modeling cancer progression and testing therapeutics in a biologically relevant context. The known and unknown essential signaling factors in the important regulation pathways would be studied for the disease monitoring and drug delivery control. We are planning to apply the present technique to the co-culture of perivascular epithelioid cells and liver cells to investigate their signaling pathway and generate a physiologically relevant *in vitro* model.

References

1. Delamarche E, Bernard A, Schmid H, Michel B, Biebuyck H (1997) Patterned delivery of immunoglobulins to surfaces using microfluidic networks. Science 276:779–781
2. Kaji H, Yokoi T, Kawashima T, Nishizawa M (2010) Directing the flow of medium in controlled cocultures of HeLa cells and human umbilical vein endothelial cells with a microfluidic device. Lab Chip 10:2374–2379
3. Phillips JA, Xu Y, Xia Z, Fan ZH, Tan WH (2009) Enrichment of cancer cells using aptamers immobilized on a microfluidic channel. Anal Chem 81:1033–1039
4. Lin BC, Liu TJ, Qin JH (2010) Carcinoma-associated fibroblasts promoted tumor spheroid invasion on a microfluidic 3D co-culture device. Lab Chip 10:1671–1677
5. Kamm RD, Chung S, Sudo R, Mack PJ, Wan CR, Vickerman V (2009) Cell migration into scaffolds under co-culture conditions in a microfluidic platform. Lab Chip 9:269–275
6. Unger MA, Chou HP, Thorsen T, Scherer A, Quake SR (2000) Monolithic microfabricated valves and pumps by multilayer soft lithography. Science 288:113–116
7. Sabban EL, Maharjan S, Nostramo R, Serova LI (2010) Divergent effects of estradiol on gene expression of catecholamine biosynthetic enzymes. Physiol Behav 99:163–168
8. Wei HB, Li HF, Mao SF, Lin JM (2011) Cell signaling analysis by mass spectrometry under coculture conditions on an integrated microfluidic device. Anal Chem 83:9306–9313

If you have any concerns about our products,
you can contact us on
ProductSafety@springernature.com

In case Publisher is established outside the EU,
the EU authorized representative is:
Springer Nature Customer Service Center GmbH
Europaplatz 3, 69115 Heidelberg, Germany

Printed by Libri Plureos GmbH
in Hamburg, Germany